HOW MANY LICKS?

Or, How to Estimate Damn Near Anything

AARON SANTOS

RUNNING PRESS
PHILADELPHIA • LONDON

For Anna

I had no idea taking the Fung Wah bus ride would be so lucky.

Printed in the United States

9 8 7 6 5 4 3 2 1
Digit on the right indicates the number of this printing

Library of Congress Control Number: 2009927928

ISBN 978-0-7624-3560-9

Design by Joshua McDonnell
Interior Illustration by Mario Zucca
Typography: Helvetica

Cover Photograph by Joshua McDonnell

Running Press Book Publishers
2300 Chestnut Street
Philadelphia, PA 19103-4371

Visit us on the web!
www.runningpress.com

CONTENTS

INTRODUCTION

WHY BE APPROXIMATE?

You're on the bus and running a little late for work. In your hurry, you forgot your watch and your cell phone is at the bottom of your briefcase, so you ask the woman next to you what time it is. She glances at her watch, which reads 8:33:46, and replies "Eight-thirty."

Did she lie? Why didn't she say, "Eight thirty-three and forty-six seconds A.M.?" Maybe she doesn't keep her watch set exactly. Maybe she, too, was in a rush and didn't want to waste precious time talking to you. Maybe she realized that by the time she finished her sentence, it would be 8:33:48.

For various reasons, people rarely ever use such precision when dealing with numbers. It probably didn't matter to you whether the time was 8:33:46 or 8:31:27, and reading off the extra digits wastes your time and hers. Whether it's saying "eight-thirty" or adding a "pinch" of nutmeg, we all approximate things. Unless you work at NASA, sacrificing a little bit of precision to save time is a good trade-off because, for the most part, we all have more important things to do. But approximation is more than just rounding to the nearest minute. It's also a valuable intellectual tool that helps us conceptualize numbers.

Approximation works as an idea filter that can be used to weed out bad ideas when making a decision. Whether you're a businessman deciding which product to pursue, a congressman voting on whether or not to build a fence around the Mexican border, or a physicist trying to detect the Higgs Boson, approximation is the first technique you should use as a feasibility test.

Let's say you're a government official working on a missile defense plan. Do you pursue it immediately, potentially wasting billions of taxpayer dollars only later to find out whether or not it ever had a chance of stopping a nuclear attack? Or should you first make a rough estimate of the plan's likelihood of success, potentially saving time and money while you come up with a legitimate plan? If you estimate the plan has a 90% chance of success, you definitely pursue it. Maybe even if you estimate it has only a 10% chance, you still pursue it. But, if your estimate concludes its chances of realization are less than the chance of winning the lottery, it would be foolish to ever consider it. Even without an advanced engineering degree, estimations like these are possible and necessary for making some of the critical decisions we face every day. Whether it be missile defense or simply when to cross the street, we make estimates all the time, and having a sound mathematical footing only improves our accuracy.

In addition to providing an idea filter, getting in the habit of formally approximating things also boosts one's numerical prowess, specifically one's ability to conceptualize very big (and very small) numbers. Although you may not yet have an appreciation for the difference between a billion and a trillion, experience using these numbers builds an understanding of them quickly. One rapidly develops "numerical landmarks" that act as a conceptual guide. For example, at the time of this writing, a billion is about one-seventh the world population and a trillion is about one-eleventh the national debt. With simple arithmetic and a little bit of practice, you can estimate just about anything (no matter how large or small) and grow comfortable understanding big numbers in the process.

How to Approximate: The Fermi Method

There are many approximation techniques, but one of the most powerful is the Fermi approximation. Its power stems from the fact that it's both easy and quick to use and can be done with little background information. Though there isn't a well-defined procedure for Fermi approximations, the general steps are to first make basic assumptions that seem reasonable and then to use those assumptions to calculate what you want to know.

Let's say I wanted to know how many leaves are on a tree. I could first estimate that each branch has about 30 leaves. For a given tree, 30 leaves is a reasonable number to have on a branch. I could then assume there are 10 branches on each tree. Some trees have more, others less, but 10 branches is a reasonable number to have on a tree. I then know that if each of the 10 branches has 30 leaves then there are (30 leaves per branch) x (10 branches per tree) = 300 leaves per tree. This example is simple enough, but for more complicated examples it helps to have some basic guidelines.

DID YOU KNOW . . .

Enrico Fermi (1901–1954) was an Italian physicist and Nobel laureate known mostly for his work on quantum theory. He was renowned for solving seemingly impossible problems using order of magnitude estimates. Half of the particles in the universe—fermions—are named after him.

1. START WITH WHAT YOU KNOW

Maybe you're calculating how much it's going to cost for all the bricks needed to build a new schoolhouse. You don't know how many bricks are in a schoolhouse, so you can't start there. Maybe you start off knowing that each brick is about half a foot long. You also know the school house is to be about 100 feet long. From this you calculate that 200 bricks are needed to make that length. Next you guess the building is to be about 20 feet high. All of these are reasonable estimates. Maybe you're off by a few percent in each one, but if you were to start by guessing the number of bricks in the whole building you could be off by a factor of 10, 100, or even more. Start with approximations that you're pretty sure about and then calculate anything you're not sure about.

2. USE UNIT CANCELLATIONS TO FIND A PATH TO THE ANSWER

In the example using leaves, I know that there are a certain number of "leaves per branch" and a certain number of "branches per tree." While you may have forgotten how you canceled units in chemistry class, it's simple enough to remember if you can see that it's just like multiplying and dividing numbers. If I take "branches per tree" and multiply by "leaves per branch," the "branches" cancel, giving "leaves per tree."

$$\frac{30\text{ leaves}}{1\text{ }\cancel{\text{branch}}} \times \frac{10\text{ }\cancel{\text{branches}}}{1\text{ tree}} = \frac{300\text{ leaves}}{1\text{ tree}}$$

It's simple factoring. One divided by one is one, 37 divided by 37 is one, branches divided by branches is one, thingy divided by thingy is one. Sometimes you need a string of units canceling to get the answer you seek. For example, in the schoolhouse calculation it may be that you use (X dollars per brick) x (Y bricks per wall) x (Z walls per schoolhouse) to get a final answer of X*Y*Z dollars per schoolhouse.

3. USE WORST-CASE SCENARIO BOUNDS

Sometimes you won't be able to start with something that's a reasonable guess. Maybe you want to know the number of teachers living in the Los Angeles city limits, but have no idea where to start. Well, you know there's certainly greater than 100 teachers, as that would fill up only two decent-sized schools. You also know that it should be less than 10% of the total city population (which in this case would be about 1 million people). Here we have an upper bound and a lower bound. Though you might not know what the answer is, you certainly know what the answer is not, and you can then use other information and the process of elimination to chip away until you find a better estimate.

4. USE THE WEB

There's no shame in using the Web to look up numbers you don't know. There are a variety of websites (Google, Wikipedia, etc.) that are useful for this purpose. Throughout the book, I will often look up things (physical constants, governmental budgets, etc.) that aren't interesting to calculate and don't aid in learning the process of approximation. Except where instructional, I won't calculate anything that can be found with a simple Google search, and neither should you.

5. BE HONEST

This is perhaps the most important rule. We all have personal biases. Especially when there's some uncertainty involved, it's important to make sure that your preconceptions about the way things "should" work do not affect the outcome of calculations you do. I've often caught myself trying to force a calculation to turn out the way I want. At times like this, it's important to use conservative estimates and the utmost skepticism when deriving a result.

6. PLAY THE GAME AND MAKE NUMERICAL LANDMARKS

The old cliché "Practice makes perfect," is certainly true when it comes to approximations. The more you do it, the better you get at it and the easier it will be to visualize numbers. Over time, you'll build landmarks in number space that'll help you navigate. For example, when you hear things like a million and a billion, you should remember that a million is about the number of seconds in a week and a half, whereas a billion seconds is a little less than 32 years. Having facts like these to refer to greatly increases one's understanding of large numbers.

Simplify Numbers

One thing that makes approximations fast and particularly easy is that you can put the numbers in a simpler form. For example, since we're only approximating, the number 397 could just as easily be rounded to 400 since the difference isn't significant given the degree of precision we're working with. Below are some tricks that I'll use to make the math simpler.

ROUNDING

Just like our fellow bus passenger, we're not interested in knowing the time exactly, only what's significant to us. For the most part, this book will deal with two significant figures. Anything that's 187 becomes 190, and anything that's 7,432 becomes 7,400. It makes the math easier and saves time.

EXPONENTIAL NOTATION

Many people prefer seeing numbers written out or hearing words like a million or a billion because "they make more sense." But can we really make more sense out of these? For example, most us would be happy to have a million or a billion dollars. With one million dollars we could live on a thousand dollars a day for a little less than three years before it ran out. How long would a billion dollars last? Six years? Thirty years? How about 2,740 years? That's right, if we had a billion dollars back in the time of Jesus, we'd still be spending it today because one billion is one thousand times greater than one million.

Our conceptions of very big numbers (and very small numbers) aren't particularly good, but exponential notation makes them effortless to think about. In exponential notation, numbers are written as a decimal number with a single non-zero digit before the decimal point times 10 to some power. The power tells us how many spaces we have to move the decimal point if we tried to write the whole number out. When you run out of places to move the decimal point, you add zeroes to the end of the number.

For example, 4.1×10^{15} could be written out by moving the decimal point one place to the right to get 41 and then adding 14 more zeroes so that you'd have a 41 followed by fourteen zeroes. The number 6.234×10^{53} could be written out by moving the decimal point 3 places to the right and then writing out 50 more zeroes after the 6234. It's easy to see that reading 6.234×10^{53} is much more comprehensible than reading

62340000000000000000000000000000000000000000000000000000.
For negative powers, you move the decimal point to the left so that the written-out number is now a zero followed by a decimal point followed by the "magnitude of the power minus one" zeroes followed by the number in front. For example, 3.23×10^{-3} is a decimal point followed by 3-1=2 zeroes followed by 323, i.e., 0.00323.

Exponential notation reduces writing space and makes it easier to read and visualize numbers. It's worth noting that in some cases it's impractical (if not impossible) to write numbers as anything but exponential notation. The largest number in the book is $3.8 \times 10^{5,767,416}$. If I were to try to write this out with the actual number of zeroes, it alone would produce a 5,000-page book!

ROUND TO AN ORDER OF MAGNITUDE

The nearest "order of magnitude" means the nearest power of 10. For example, 0.00001, 1, and 1,000,000 are all integer powers of 10. Written in exponential notation, I would have 10^{-5}, 100, and 10^{6}, respectively. Using a property of exponentials, doing calculations with these numbers becomes very easy because you can multiply by adding the exponents. For example: $10^{2} \times 10^{5} = 10^{7}$ and $10^{2} \times 10^{-5} = 10^{-3}$.

About the Format

The problems in the book are roughly organized from easy to difficult. The first several problems serve as an introduction to the Fermi approximation technique and lay out the format for the rest of the book. These problems are less intimidating than later problems, and several have answers that can be looked up. In this way, they demonstrate the power of the Fermi approximation by showing how accurate it can be.

Problems in the middle of the book are more likely to require multiple steps and may deal with mathematical and physical concepts that the reader is less familiar with.

Finally, the last set of problems deals with scientific concepts and algebraic equations that may be quite daunting for the unfamiliar reader. These are the problems that you do for honor as well as enjoyment.

Each problem consists of six parts:

- First, **A BACKGROUND STORY** describes the question we're considering and gives my own take on the problem. In some instances, the background story contains numbers that will help solve the problem.
- **ASK YOURSELF THIS . . .** —This is a list of questions that provide some useful ideas you might want to consider before trying to solve the problem.
- **HELPFUL HINTS**—I address each of the questions asked above and list the numbers I use to calculate the answer. If the problem requires use of physics equations, a brief tutorial is given on the physics that underlies the problem. (Feel free to skip this section if you're feeling studly and want to solve the problem on your own!)
- **CONSTRUCT A FORMULA**—I describe in words the operations you can do to get the result. This is a good place to see if your units have canceled the right way.
- **MESSY MATH**—The numbers are plugged in.
- **ANSWERS**—Finally, I show the results and try to provide some context.

One Last Thing

One disclaimer before we dive in: This is *not* a book of answers. I make no guarantees that the answers are right. This is a book about methods, about how to use numbers to calculate things. Don't buy this book if you're only looking to find out some interesting factoids[i]. I won't be happy if you buy it and don't learn something from it[ii]. If this book accomplishes its task, then by the end you should be able to approximate answers to questions that you come up with. In fact, you'll probably even be able to do a better job on some of the things I've estimated in the book. After all, there's more than one way to skin a math problem. Ultimately, if you end up doing a better job than what I've written, than I'll feel like my work here is done.

[i] On second thought, buy the book even if you're only looking to find some interesting factoids. Far be it for me to say who should or shouldn't pay me money. Just don't blame me if the numbers aren't quite right, because they're just approximations.

[ii] I'll still be happy if you buy the book. I'll just be happier if it helps you learn how to enjoy and appreciate numbers.

1. Fine Tuning

To illustrate the power of the methods described in the previous section, let's start with Fermi's classic example: **How many piano tuners are in Chicago?** Seem impossible? At first glance, such questions may seem hopeless to answer without looking in the Chicago Yellow Pages. As you'll soon see, it's not nearly as difficult as it seems.

We have to start somewhere, so let's start with what we know. How many people live in Chicago? Okay, we probably don't know that, but maybe we know how many people live in Boston or Detroit or any other major city. Most major cities have around 3.0×10^6 (3 million) people living in them. Some have more, others less, but they're usually around that size. If it's not exact, it should be close. Even in the worst-case scenario, it will certainly be between 500,000 (a medium-sized city) and 50 million (more than the largest city in the world). We have to make basic assumptions like these, but we shouldn't sweat how precise they are since they're just approximations that give us *roughly* the right number. If the calculation needs to be really precise, you can always go online and look up the exact result (2.8 million people, according to Wikipedia at the time of this writing), but if you like results fast and dirty, this is the way to go.

Moving on, we'll next guess that 30 percent of residents are homeowners[iii]. We'll assume that 5 percent of these homeowners have pianos and that their pianos are probably tuned once every six months. We don't consider the number of musical halls, performance areas, etc. because they're probably much fewer than the number of nonprofessionals who own pianos. Lastly, a piano tuner would probably tune about two pianos a day, or equivalently 730 pianos a year.

[iii] Some of the numbers used in this book may seem arbitrary or even downright wrong. Before you send me hate mail, remember that they're an order of magnitude estimate. Ask yourself whether the number is within a factor of 10. For example, the percentage of residents who are homeowners may not be exactly 30 percent, but it's certainly between 3 percent and 300 percent!

ASK YOURSELF THIS...

A) How many people are in Chicago?

B) What percentage of them are homeowners?

C) What percentage own pianos?

HELPFUL HINT:

There are about 3.0×10^6 people in Chicago and other major cities.

CONSTRUCT A FORMULA:

Here, we construct a formula that can be used to find the number of pianos tuners. We do this by combining the variables listed above in such a way that their units cancel leaving us with units of "piano tuners." In order to simplify our formula, we can do a few simple calculations first.

1) For example, if we multiply the fraction[iv] of residents who are homeowners (0.3 homeowners per resident) times the number of total residents (3.0×10^6 residents) we get the number of homeowners (9.0×10^5 homeowners).

2) If we multiply this again by the fraction of homeowners who own pianos (0.05 piano owners per homeowner), we find there are about 4.5×10^4 pianos.

3) These pianos need to be tuned twice each year, meaning that there are about 9.0×10^4 tunings per year.

4) Finally, we can obtain the number of piano tuners by dividing the number of tunings per year by the number of tunings per year per tuner (730 tunings per year per tuner):

$$\frac{\text{(total \# of tunings per year)}}{\text{(\# of tunings per year per tuner)}}$$

[iv] If you haven't worked with percentages in a while, remember that you must divide a percentage by 100 to get the same number as a fraction and multiply a fraction by 100 to get the same number as a percentage. For example, 30 percent is equivalent to 0.3.

MESSY MATH:

$$\frac{\left(9.0 \times 10^4 \,\frac{\text{tunings}}{\text{year}}\right)}{\left(730 \,\frac{\text{tunings}}{\text{year} \cdot \text{tuner}}\right)} \approx 120 \text{ piano tuners}$$

ANSWER:

This is a reasonable estimate. The Chicago Yellow Pages has 43 listings under piano tuners and not every tuner pays for Yellow Pages listings, so we're off by less than a factor of 3.

While calculating the number of piano tuners may not seem particularly interesting, it's quite extraordinary when you consider the technique gives similar accuracy for the number of stars in the sky or grains of sand on the Earth or various other quantities. Here we have an easy way to answer questions that at first glace seem impossible!

2. Populations: Cities, States, Countries, and World

Many of the calculations in this book use populations. For that reason, it's good to know how to estimate populations of cities, states, and countries[v]. First, it's helpful to remember that a very large city, like New York, has about 8 million = 8.0×10^6 people living in it. A smaller city, like Des Moines, has about 200,000 = 2.0×10^5 people. On average, a city has about 500,000 = 5.0×10^5 people in it. **HOW MANY PEOPLE ARE IN THE WORLD?**

ASK YOURSELF THIS...

A) What fraction of a state's population lives in its largest city?
B) What fraction of a U.S. population lives in an average-sized state?
C) What fraction of the world population lives in the United States?

HELPFUL HINTS:

- A large city makes up about 10% of a state's population. This means there are 10 state residents for every one resident of a large city.
- Since there are 50 states in the United States, an average-sized state must make up 1/50th = 0.02 = 2% of the United States population. Alternatively, you can say there are 50 U.S. residents per state resident.

[v] In the back of the book, you can look up useful numbers such as the world population, the mass of Saturn, the distance to the Sun, etc.

- The United States has about 5 percent of the world population. Equivalently, you can say there are 0.05 U.S. residents per world resident. You can figure this out by assuming that people are distributed fairly uniformly across the world and by approximating the United States as being about 5 percent of the total land area on Earth.

CONSTRUCT A FORMULA:

1) Multiply the number of people in a large city (5.0×10^5 people) times the ratio of state to city residents (10 state residents per city resident) to get the number of people per state.
2) Next, multiply this by the ratio of U.S. residents to state residents (50 U.S. residents per state resident) to get the population of the United States (2.5×10^7 U.S. residents).
3) Finally, divide this by the fraction of the world population that resides in the United States (0.05 U.S. residents per world resident) to obtain the world population. Combine these steps into one formula, and you get the following:

$$\frac{(\text{\# of city folk}) \times (\text{ratio of state to city folk}) \times (\text{ratio of U.S. to state folk})}{(\text{ratio of U.S. to world folk})}$$

MESSY MATH:

$$\frac{(5.0 \times 10^5 \text{ city folk}) \times \left(10 \dfrac{\text{state folk}}{\text{city folk}}\right) \times \left(50 \dfrac{\text{U.S. folk}}{\text{state folk}}\right)}{\left(0.05 \dfrac{\text{U.S. folk}}{\text{world folk}}\right)} \approx 5.0 \times 10^9 \text{ peeps}$$

ANSWER:

We predict about 5 billion people in the world. At the time of this writing, a Google search shows a world population of 6.7 billion, so we're only off by 35 percent, which is still a reasonable estimate.

We also calculated the populations of an average state (5.0×10^6 people) and of the United States (2.5×10^8 people). These are pretty good approximations when you consider that Colorado—the 22nd most populous state—has 4.9×10^6 people and that the United States has 3.0×10^8 million people.[vi]

[vi] These numbers were obtained from the 2008 estimates of the U.S. Census Bureau as listed in Wikipedia under the entry "List of U.S. states by population."

3. Conversions: Smoots, Bols, and Gaedels

Being able to convert between different types of units is essential for estimating things. While most of the world uses the superior metric system, Americans are still stuck measuring distances with a dead king's foot. (Seriously, I thought we kicked Britain's royal bottom in 1776 . . . Why are we still doing this?!) Because it's easier to convert things, I'll use metric units for most of the problems in this book and include English units in parentheses when possible.[vii] Aside from convenience, the choice of units is essentially arbitrary. To illustrate this point and provide some explanation on how to convert between different units, let's consider the following case.

Anyone who has crossed the Harvard Bridge in Cambridge during a cold wintry night knows that the trek can seem endless. In October 1958, an MIT fraternity decided that it was time that the bridge should be labeled so that travelers could know how far they'd gone. Oliver Reed Smoot Jr., a pledge at the fraternity, was chosen to be the 1.7 m (~5 ft 7 in) measuring stick that would mark off the bridge. His frat brothers picked him up and laid him down continuously until they had reached the end of the bridge, which they determined was a total distance of 364.4 Smoots plus

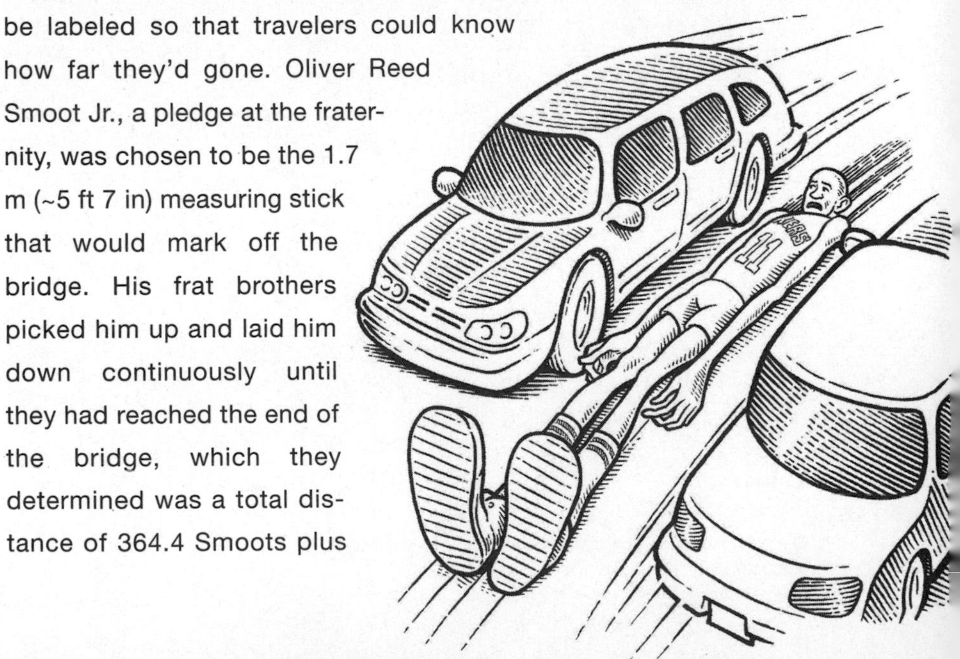

one ear. What if the fraternity members had used someone else like the 1.1 m (~3 ft 7 in) former MLB attraction Eddie Gaedel or the 2.3 m (~7 ft 6in) former NBA center Manute Bol? **WHAT IS THE LENGTH OF THE HARVARD BRIDGE MEASURED IN GAEDELS? IN BOLS?**

ASK YOURSELF THIS...

A) How many Gaedels are there per Smoot?

B) What about Bols per Smoot?

HELPFUL HINT:

The length of the bridge is 364.4 Smoots plus one ear. Assume you can neglect the ear since it's small compared to the rest of the bridge.

CONSTRUCT A FORMULA:

1) Divide the height of Smoot (1.7 m per Smoot) by the height of Gaedel (1.1 m per Gaedel) to get the number of Gaedels per Smoot (1.5 Gaedels per Smoot).
2) Likewise, divide the height of Smoot (1.7 m per Smoot) by the height of Bol (2.3 m per Bol) to get the number of Bols per Smoot (0.74 Bols per Smoot).
3) Multiply these conversion factors by the length of the bridge (364.4 Smoots) to get the bridge length in the appropriate units:

(# of Smoots in the bridge)
x (conversion factor from Smoots to Gaedel or Bols)

[vii] Many of the conversion factors that you need to change between units are in the back of the book.

MESSY MATH:

For Gaedels:

$$\left(364.4 \text{ Smoots}\right) \times \left(1.5 \ \frac{\text{Gaedels}}{\text{Smoot}}\right) \approx 546.6 \text{ Gaedels}$$

or

For Bols:

$$\left(364.4 \text{ Smoots}\right) \times \left(0.74 \ \frac{\text{Bols}}{\text{Smoot}}\right) \approx 269.7 \text{ Bols}$$

ANSWER:

The Harvard Bridge is about 546.6 Gaedels or, equivalently, 269.7 Bols.

4. Visualizing Large Numbers

This book deals with a lot of large numbers, and visualization is an important part of conceptualizing them. Let's practice.

In high-school chemistry, we all learned about *moles*. A mole is just a number like "a dozen," except instead of 12, it's ~602,200,000,000,000,000,000,000 = 6.022×10^{23}. It's a big number, roughly the number of atoms in objects that are big enough for us to notice. Imagine a mole of chocolate jelly doughnuts melting in the hot sticky August heat. **WHAT FRACTION OF EARTH'S VOLUME WOULD A MOLE OF HOT STICKY CHOCOLATE JELLY DOUGHNUTS BE?**

ASK YOURSELF THIS...

A) How big is a jelly doughnut?

B) What would a line of 1 million jelly doughnuts look like? What about a square of 1 million by 1 million jelly doughnuts? What about a cube of 1 million by 1 million by 1 million jelly doughnuts?

C) How many of these doughnut cubes are needed to make a mole?

D) What's the volume of the Earth?

HELPFUL HINTS:

- Doughnuts are about 10 cm = 0.1 m (~4 in) from one end to the other.
- Imagine 1 million doughnuts lined up in a row, slowly decaying in the sun. By multiplying 1 million times the size of a doughnut, we can find the total length of this row of doughnuts to be 100 km (~62 mi). This would stretch across the state of Rhode Island.

We've just eliminated a factor of 1 million doughnuts, so we now have 602,200,000,000,000,000,~~000,000~~ left. Now imagine making a square 1 million doughnuts long and 1 million doughnuts wide covering the entire state of Rhode Island. We've now eliminated another factor of 1 million doughnuts, so we have 602,200,000,000,~~000,000,000,000~~ doughnuts left. Let's now stack the gooey doughnuts on top of each other. It's now a cube of doughnuts sitting right on top of Rhode Island. We've now shaved off another factor of 1 million, leaving 602,200,~~000,000,000,000,000,000~~ doughnuts left.

- In *Appendix C*, you can look up the radius of the Earth. You can then calculate Earth's volume using the formula for the volume of a sphere: $V = 4 \pi r^3 / 3$, where r is the radius and π (pronounced "pi") is equal to 3.1415926... Using this formula, the volume of the Earth is equal to 1.1×10^{12} km^3.

CONSTRUCT A FORMULA:

1) By cubing the length, we can find the volume per cube (1.0×10^6 km^3 per cube).
2) Multiply the # of these cubes needed to make a mole (602,200 cubes per mole) by the volume per cube to obtain the total volume of a mole of jelly doughnuts.
3) Divide this by the volume of the Earth to obtain the fraction of Earth's volume that the cube would take up:

$$\frac{\text{(\# of cubes) x (volume per cube)}}{\text{(volume of the Earth)}}$$

MESSY MATH:

$$\frac{\left(6.022 \times 10^{5}\,\frac{\text{cubes}}{\text{mole}}\right) \times \left(1.0 \times 10^{6}\,\frac{\text{km}^{3}}{\text{cube}}\right)}{\left(1.1 \times 10^{12}\,\frac{\text{km}^{3}}{\text{Earth}}\right)} \approx 0.54 \text{ Earths per mole}$$

ANSWER:

A mole of jelly doughnuts would be about half the size of the Earth. This is larger than the Moon.

5. Agony of Da Feet

It has been said that you have to "walk a mile in a man's shoes before you can judge him." **HOW MANY MILES DOES A PERSON WALK IN A LIFETIME?**

ASK YOURSELF THIS...

A) How long does a person live?

B) What percentage of a person's day is spent walking?

C) How fast does a person walk?

HELPFUL HINTS:

- A ripe old age is about 80 years = 7.0×10^5 hours.
- The amount of time spent walking each day varies considerably from person to person based on age, physical fitness, career choice, etc. An average person gets about eight hours of sleep during which he or she is presumably not walking. Much of the other 16 hours of the day are spent sitting, lying, or standing but not actually walking. Since we almost certainly walk more than 0.5 percent of the time but less then 50 percent of the time, assume 5 percent of our waking hours are spent walking.
- A person can leisurely walk about 1.0 mile in an hour.

CONSTRUCT A FORMULA:

1) Multiply the total number of hours in a lifetime (7.0×10^5 total hours per lifetime) times the fraction of the day spent awake (0.67 waking hours per total hour) to obtain the total number of waking hours in a lifetime.
2) Multiply this by the fraction of waking hours spent walking (0.05 walking hours per waking hour) to get the total number of hours walked in a lifetime.
3) Finally, multiply this times the walking speed (1.0 mile per hour) to obtain the number of miles walked per lifetime:

(# of total hours per lifetime) x (# of waking hours per total hour) x (# of walking hours per waking hour) x (# of miles per hour)

MESSY MATH:

$$\left(7.0 \times 10^5 \frac{\text{total hrs}}{\text{lifetime}}\right) \times \left(0.67 \frac{\text{waking hrs}}{\text{total hrs}}\right) \times \left(0.05 \frac{\text{walking hrs}}{\text{waking hrs}}\right)$$

$$\times \left(1.0 \frac{\text{mile}}{\text{hour}}\right) \approx 23{,}000 \text{ miles per lifetime}$$

ANSWER:

You'll walk about 23,000 miles in a lifetime. That's long enough to walk around the Earth.

6. A Hairy Situation

A Vietnamese man hasn't visited a barber in over 30 years in an attempt to get into *The Guinness Book of World Records*. Tran Van Hay's hair has grown twenty feet in that time. **How long does a person's hair typically grow in a lifetime?**

ASK YOURSELF THIS...

A) How long does your hair grow in a month?

B) How many months will you live?

HELPFUL HINTS:

- Although it varies from person to person, hair tends to grow about 1.0 cm (~0.39 in) per month. If you keep the length of your hair the same and get your hair cut once per month, this is the length that you would cut off each time.
- If you're reasonably healthy, you'll likely live about 80 years = 960 months.

CONSTRUCT A FORMULA:

Multiply the length of hair grown per month (1.0 cm per month) times the number of months per lifetime (960 months per lifetime) to get the total length of hair grown in a lifetime:

(length grown per month) x (# of months per lifetime)

MESSY MATH:

$$\left(1\ \frac{\text{cm}}{\text{month}}\right) \times \left(960\ \frac{\text{months}}{\text{lifetime}}\right) \approx 960 \text{ cm per lifetime}$$

ANSWER:

After 80 years with no barber, your hair would be about 960 cm = 9.6 m (~31 ft) long.

If we tied all of this head hair together end-to-end to form a rope, it would stretch almost 2,400 km, long enough to reach halfway across the continental United States.

7. Wolverine Who?

World records always intrigue people: the world's longest hair, the world's tallest woman, the world's fattest twins, etc. One such interesting record was set by Shridhar Chillal of India. According to *The Guinness Book of World Records*, Chillal has the longest fingernails on a single hand, the total measurement of the five nails on his left hand being 7.05 m (~23 ft 1.5 in). He last cut his nails in 1952. **WHAT IS THE TOTAL LENGTH OF FINGERNAIL GROWTH THAT ONE COULD ACHIEVE IN A LIFETIME?**

ASK YOURSELF THIS...

A) How long do your nails grow between clippings?

B) How often do you clip your nails?

C) How long does a person live?

HELPFUL HINTS:

- The speed at which fingernails grow varies from person to person, but they typically grow about 2 mm = 0.002 m (~0.078 in) before you cut them.
- People usually cut their nails about once a week.
- There are 4,160 weeks in an 80-year lifetime.

CONSTRUCT A FORMULA:

Multiply the length grown per clipping (0.002 m per clipping) times the number of clippings per week (1 clipping per week) times the number of weeks per lifetime (4,160 weeks per lifetime) to get the total length of nail growth per lifetime:

(length grown per clipping) x (# of clippings per week) x (# of weeks per lifetime)

MESSY MATH:

$$\left(\frac{0.002\text{ m}}{1\text{ clipping}}\right) \times \left(\frac{1\text{ clipping}}{1\text{ week}}\right) \times \left(\frac{4160\text{ weeks}}{1\text{ lifetime}}\right) \approx 8.3\text{ m per lifetime}$$

ANSWER:

You can grow about 8.3 m (~27 ft) of nails. Just about what you'd expect from Chillal's nails.

8. You Piece of Shoe!

Here's a calculation that's good for the sole. It's one that'll really make your tongue hang out. It's really a work of boot-y—hey, wait! Where you going? I was just getting started. Oh, well. **How many times can you outline the continental United States in shoelaces?**

ASK YOURSELF THIS...

A) How many people are in the United States?

B) How many shoelaces does an average American own?

C) How long is each lace?

D) What's the length of the perimeter of the continental United States?

HELPFUL HINTS:

- In problem 2, we showed how to calculate populations. There are about 3.0×10^8 people in the United States.
- Some Americans have dozens of shoes, but most of us have significantly fewer pairs. For simplicity, assume that an average person owns three pairs of (lace-able) shoes. Given two shoelaces per pair, that's six laces per person.
- There's about 1.0 m (~3.3 ft) per lace.
- You can estimate the perimeter length in various ways. For example, you can approximate it by using a ruler and a U.S. map with a scale bar[viii].
- Alternatively, if you do a Google search for the phrase "perime-

[viii] Technically, the length of a coastline is difficult to measure due to its "fractal" nature. Fractals have the peculiar property that their length depends on the ruler with which you measure them. The curious reader can find more information about this by running a Google search for "The Richardson Effect."

ter of the U.S.," you come across a bike tour that travels around the continental United States in 2.1×10^7 m = 2.1×10^4 km (~1.3×10^4 mi).

CONSTRUCT A FORMULA:

1) Multiply the number of people in the United States (3.0×10^8 people) times the number of laces per person (6 laces per person) times the length per lace (1.0 m per lace) to get the total length of all the laces.
2) Divide this result by the length of the U.S. perimeter (2.1×10^7 m) to obtain the number of times you could encircle the continental United States in American shoelaces:

$$\frac{\textbf{(\# of people) x (\# of laces per person) x (length per lace)}}{\textbf{(perimeter length of the United States)}}$$

MESSY MATH:

$$\frac{(3.0 \times 10^8 \text{ people}) \times \left(6 \frac{\text{laces}}{\text{person}}\right) \times \left(1.0 \frac{\text{m}}{\text{lace}}\right)}{(2.1 \times 10^7 \text{ m})} \approx 86 \text{ times}$$

ANSWER:

You could outline the United States in shoelaces about 86 times.

9. Hammerin' Hank

Major League Baseball's steroid scandals during the late 1990s and early 21st century have tainted many of the records broken during that span. While these scandals have caused fans to look with suspicion on any record set during this era, they also put in perspective how remarkable Hank Aaron's homerun total is. With a little math, it's possible to provide even more perspective. **What's the total distance of Hammerin' Hank's 755 long balls and the total mileage of his homerun trots?** *(Since Major League baseball diamonds are usually measured in feet, keep this in English units.)*

ASK YOURSELF THIS...

A) How far does a homerun travel?

B) What's the distance around the bases?

HELPFUL HINTS:

- According to *The Guinness Book of World Records*, the longest homerun measured in an MLB game is 634 feet by Mickey Mantle of the New York Yankees. This is an extreme example, however, since more often than not, homeruns just barely make it over the wall. The distance to the outfield wall varies from park to park but generally lies within a range of 300 to 400 feet. Assume the average homerun travels 400 feet.
- There's exactly 90 feet between each base, so there's 360 feet around the bases.

CONSTRUCT A FORMULA:

1) Multiply the number of homeruns (755 homeruns) by either the number of feet per homerun (400 ft per homeruns) or the number of feet per homerun trot (360 per homerun trot).

2) Multiply these results by the conversion factor from feet to miles (1 mi per 5,280 ft) to calculate the answer in miles:

(# of homeruns) x (# of feet per homerun) x (# of miles per foot)

or

(# of homeruns trots) x (# of feet per homerun trot)
x (# of miles per foot)

MESSY MATH:

For the homeruns:

$$\left(755 \text{ homeruns}\right) \times \left(400 \frac{\text{ft}}{\text{homeruns}}\right) \times \left(1 \frac{\text{mile}}{5280 \text{ ft}}\right) \approx 57 \text{ miles of homeruns}$$

or

For the trots:

$$\left(755 \text{ homeruns}\right) \times \left(360 \frac{\text{ft}}{\text{homeruns}}\right) \times \left(1 \frac{\text{mile}}{5280 \text{ ft}}\right) \approx 51 \text{ miles of trots}$$

ANSWER:

Hank Aaron had about 57 miles of homeruns and 51 miles of homerun trots. That's right: Hank Aaron ran almost two marathons' worth of homerun laps.

10. Book 'em, Dano

Have you ever taken a class you were completely lost in? Did it feel like there must be some introductory course you were expected to have taken, but they stopped offering it a long time ago and just forgot to replace it? Many students experience this at some point. Some even go through a short-lived ambitious period where they attempt to read every book in their field that the library has to offer, starting with the A's and working their way to the Z's. **How long would it take to read every book in the library?**

ASK YOURSELF THIS...

A) How many books are in the library?

B) How many books can you read in a day?

HELPFUL HINTS:

- The United States Library of Congress in Washington, D.C. holds over 118 million items (requiring more than 500 miles of shelving!).
- The time it takes to read a book clearly depends on whether the book is *The Cat in the Hat* or *Finnegans Wake*. Fortunately, most books lie somewhere in between. Assume that a person can read two books per day on average or, equivalently, that the time it takes to read a book is half of a day.

CONSTRUCT A FORMULA:

Multiply the number of books in the library (1.18×10^8 books) by the time spent per book (0.5 days per book) to calculate the total time needed to read every book in the library:

(# of books in the library) x (# of days per book)

MESSY MATH:

$$\left(1.18 \times 10^8 \text{ books}\right) \times \left(0.5 \frac{\text{days}}{\text{book}}\right) \approx 5.9 \times 10^7 \text{ days}$$

ANSWER:

It would take 59 million days or 160,000 years to read every book in the library. That's almost as long as humans have been on the planet.

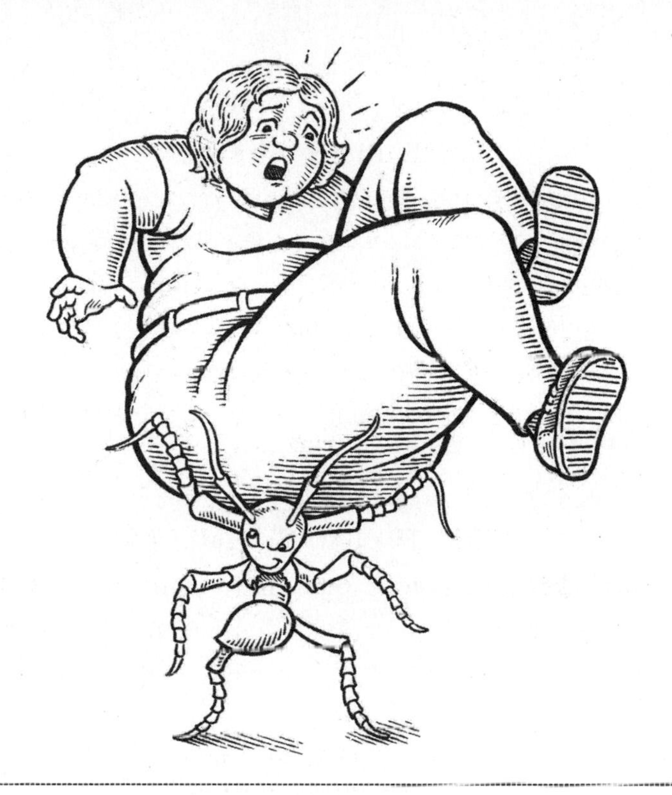

11. Ant Skates

Chitin, the material that makes up an insect's exoskeleton, is used in many practical applications and products including water purification, food additives, dyes, surgical thread, and fertilizer. This remarkable material is the reason why it's said that ants can hold 50 times their own weight. If this were true, **HOW MANY ANTS WOULD IT TAKE TO HOLD UP A HUMAN?**

ASK YOURSELF THIS...

A) How much does an ant weigh?

B) How much weight can an ant hold?

C) How much does a person weigh?

HELPFUL HINTS:

- There are over 12,000 species of ants known today spanning a wide range of sizes. According to the Wikipedia entry for "Ant," even within a species there can be a 500-fold difference in their dry-weights. For simplicity, assume ants weigh about 20 mg, so that they can hold about 1.0 g (~0.035 oz).
- A typical person weighs about 65 kg = 6.5×10^4 g (~140 lbs).

CONSTRUCT A FORMULA:

Divide the weight of a person (6.5×10^4g) by the amount of weight held per ant (1.0 g per ant) to obtain the number of ants needed to hold up a person:

$$\frac{\text{(weight of a person)}}{\text{(weight held per ant)}}$$

MESSY MATH:

$$\frac{\left(65{,}000 \frac{\text{g}}{\text{person}}\right)}{\left(1 \frac{\text{g}}{\text{ant}}\right)} \approx 6.5 \times 10^4 \text{ ants per person}$$

ANSWER:

You need 65,000 ants to hold up a human.

12. House Party

The 1990s hip-hop duo Kid 'n Play rocked the party right in their New Line Cinema classic *House Party*. I will never forget Kid's six-inch high top fade no matter how many years of therapy I pay for. **If we were to throw a house party and invite the entire world, how big of a house would we need?**

ASK YOURSELF THIS...

A) How much room does each person take up?

B) What's the world population?

HELPFUL HINTS:

- To fit everyone reasonably comfortably, you need to make sure there's enough space for each person. To help visualize, imagine shipping yourself in a crate. If you're not too tall, you could probably fit inside a box that's 2.0 m x 1.0 m x 1.0 m (~6.6 ft x 3.3 ft x 3.3 ft). However, if you're tall, portly, or claustrophobic, it might be good to have some extra space, so let's assume the box dimensions are 2.0 m x 2.0 m x 1.0 m (~6.6 ft x 6.6 ft x 3.3 ft). This means you need a total volume of 4.0 m^3 (~140 ft^3) per person.
- We discussed how to calculate populations in the second problem of this book. At the time of this writing, the world population is about 6.7×10^9 people.

CONSTRUCT A FORMULA:

Multiply the volume needed per person (4.0 m^3 per person) times the number of people (6.7×10^9 people) to obtain the total volume needed:

(volume needed per person) x (# of people)

MESSY MATH:

$$\left(\frac{4.0\ m^3}{1\ person}\right) \times \left(6.7 \times 10^9\ people\right) \approx 2.7 \times 10^{10}\ m^3$$

ANSWER:

You'd need 2.7×10^{10} m^3. At around ~10^{13} m^3 the Grand Canyon would easily provide more than enough space for the world's largest house party!

13. To Bathe or Not to Bathe...

With greenhouse gases grabbing most of the eco-threat headlines lately, people often forget that water shortage remains a big problem in many areas of the world. Whether or not water is wasted more on showers or baths clearly depends on how long of a shower you take and how big your tub is. **How long can you shower and still make it more environmentally friendly than taking a bath?**

ASK YOURSELF THIS...

A) What is the flow rate of water in a showerhead?

B) How much water is used in a bathtub?

HELPFUL HINTS:

- The rate of water flow depends on the showerhead. Older models conserve less with a flow rate of 19-30 L (~5-8 gal) of water per minute, while newer models give off as little as 3.0 L (~0.8 gal) per minute.
- Bathtubs typically have dimensions of roughly 1.5 m x 0.66 m x 0.5 m (~4.9 ft x 2.2 ft x 1.6 ft), giving a total volume of about 0.5 m^3. If we assume a normal bath is 50% full before a person steps in, then there's 0.25 m^3 = 250 L (~66 gal) of water being used per bath.

CONSTRUCT A FORMULA:

Divide the volume of water used in a bathtub (250 L) by the volume of water used per minute in either the old (30 L per min) or new (3.0 L per min) showerheads to obtain the time at which the shower starts to use more water than the bath:

$$\frac{(\text{\# of L})}{(\text{\# of L per min})}$$

MESSY MATH:

For old showerheads:

$$\frac{(250\ \text{L})}{\left(30\ \frac{\text{L}}{\text{min}}\right)} \approx 8.3\ \text{min}$$

or

For new showerheads:

$$\frac{(250\ \text{L})}{\left(3.0\ \frac{\text{L}}{\text{min}}\right)} \approx 83\ \text{min}$$

ANSWER:

That means there's a cutoff time of about eight minutes for an old model or eighty minutes for a new ultra low flow showerhead. More time than this and a bath conserves more water; less time than this a shower conserves more water.

14. Hair Today, Gone Tomorrow

Anyone who has tried growing long hair may have noticed a disturbing trend: your hair seems to start falling out more. All of sudden, everywhere you go, there's hair. In the shower drain. In the sink. On the bed. Hair, hair, hair, hair, hair! Does it mean you're going bald, or is the hair just more noticeable now that it's long? Instead of looking obsessively in the mirror and shampooing very, very gently, try estimating how much hair you have to waste. **HOW MANY HAIRS DOES A PERSON HAVE ON HIS OR HER HEAD?**

ASK YOURSELF THIS...

A) What's the density of hair on a person's head?

B) How much area is covered by hair?

HELPFUL HINTS:

- If you stare closely in the mirror, you can count the number of hairs in a row. There are about 20 hairs in a row for every centimeter of length, or roughly 400 hairs per cm^2.
- A human head is roughly spherical with a radius of about 10 cm (~3.9 in). Way back in high-school geometry, we learned that the area of a sphere is given by the equation $A = 4\pi r^2$ where r is the radius and π (pronounced "pi") equals 3.1415926... The part of a person's head covered by hair is only about half of the total area. This means the area of the hairy part is roughly $2\pi r^2$. This gives a total area of 630 cm^2 (~98 in^2) that is covered by hair.

CONSTRUCT A FORMULA:

Multiply the density of hair (400 hairs per cm^2) by the total area covered by hair (630 cm^2) to calculate the total number of hairs:

(density of hair) x (total area of hair)

MESSY MATH:

$$\left(400\,\frac{\text{hairs}}{\text{cm}^2}\right) \times (630\ \text{cm}^2) \approx 2.5 \times 10^5\ \text{hairs}$$

ANSWER:

There are about 250,000 hairs on a head, so you have quite a bit of room to play with. If you lose only 10 hairs a day, it would take almost 70 years to lose it all.

15. The Subway Shuttle

NASA's space missions are expensive. Consider NASA's New Horizons probe, which left on an Atlas 5 rocket for a nine-year journey to Pluto. The total mission cost was about $650 million, but this is a difficult number for many to comprehend. Let's put it into perspective. Most people consider subway fare fairly cheap. **WHAT COSTS MORE PER MILE: THE PROBE OR THE NEW YORK SUBWAY?**

ASK YOURSELF THIS...

A) What's the distance to Pluto?

B) How far do you travel on the subway?

C) What's the cost of a subway ride?

HELPFUL HINTS:

- In *Appendix C,* you can look up the radius of planetary orbits. The average orbital radius of Pluto is 5.9×10^9 km (~3.7×10^9 mi). The Earth goes around the Sun much faster than Pluto does (one year on Pluto is 248 Earth years), so the distance between Earth and Pluto varies, but on average it should be about the size of Pluto's orbital radius.
- Two close subway stops can be as little as 0.15 km (~0.1 mi) apart.
- No matter how far you go, a subway ride typically costs around $2.

CONSTRUCT A FORMULA:

1) To get the cost per mile for the probe, divide the cost of the NASA mission ($650 million) by the distance to Pluto (5.9×10^9 km).
2) Likewise, divide the cost of a subway ride ($2.00) by the distance traveled (0.15 km):

$$\frac{\text{(cost of trip)}}{\text{(distance traveled)}}$$

MESSY MATH:

Cost per mile on the NASA probe: $\left(\frac{\$6.5 \times 10^8}{5.9 \times 10^9 \text{ km}}\right) \approx \$0.11$ per km

or

Cost per mile on the subway: $\left(\frac{\$2}{0.15 \text{ km}}\right) \approx \$13$ per km

ANSWER:

This means that the cost of riding the subway one stop can be about 100 times more expensive per mile than riding a NASA probe!

16. The Coin Jumped over the Moon

The previous example shows how, when you think about it, NASA missions aren't that expensive. This example shows how, when you think about it, NASA missions are extremely expensive. Consider the Apollo program that led to the lunar landing. This program cost $25 billion[ix]. **IF WE WERE TO STACK UP THE EQUIVALENT AMOUNT IN PENNIES, HOW HIGH WOULD IT REACH?**

ASK YOURSELF THIS...

A) How thick is a penny?

B) What's the distance to the Moon?

C) How many pennies are in $25 billion?

HELPFUL HINTS:

- A stack of 10 pennies is about 1.5 cm (~0.78 in) high, so one penny is about 0.15 cm (~0.059 in) thick.
- From *Appendix C,* the mean distance to the Moon is 3.8×10^{8} m $= 3.8 \times 10^{10}$ cm (~240,000 mi).
- There are 2.5×10^{12} pennies in $25 billion.

[ix] Correcting for inflation, the cost would be closer to $140 billion today.

CONSTRUCT A FORMULA:

Multiply the thickness per penny (0.15 cm per penny) times the number of pennies (2.5 x 10^{12} pennies) to get the total height of pennies:

(thickness per penny) x (# pennies)

MESSY MATH:

$$\left(0.15 \frac{\text{cm}}{\text{penny}}\right) \times (2.5 \times 10^{12} \text{ pennies}) \approx 3.8 \times 10^{11} \text{ cm}$$

ANSWER:

They would reach over 3.8 x 10^{11} cm or 3.8 million kilometers. This could reach the Moon 10 times!

17. Walking on the Sun

Everyone's had awful vacations: children complaining, Dad obsessively taking pictures, Mom lecturing on how getting there is half the fun, and of course, the hours and hours of being stuck in a packed car all for a few hours on a humid beach in the baking sun. But if you think the blistering drive to Florida took forever, just imagine how long a trip to the actual Sun would be. **How long would it take to drive to the Sun?**

ASK YOURSELF THIS...

A) What's the distance to the Sun?

B) How fast does Dad drive?

HELPFUL HINTS:

- From *Appendix C,* the distance from the Earth to the Sun is about 1.5×10^{11} m (~ 9.3×10^{7} mi).
- Assuming Dad stays within the speed limit, the car will travel at a speed of about 27 m/s (~60 mph).

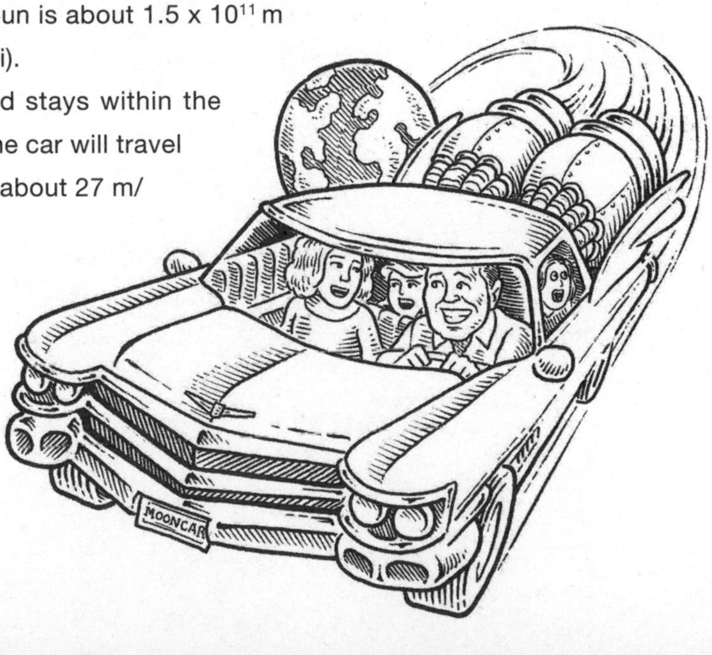

CONSTRUCT A FORMULA:

Divide the distance to the Sun (1.5×10^{11} m) by the speed of the car (27 m/ s) to obtain the total time it takes to drive to the Sun:

$$\frac{\text{(distance)}}{\text{(speed)}}$$

MESSY MATH:

$$\frac{(1.5 \times 10^{11}\ \text{m})}{(27\ \text{m/s})} \approx 5.6 \times 10^{9}\ \text{s}$$

ANSWER:

It would take about 180 years to drive to the Sun. In contrast, it takes seven minutes for light traveling from the Sun to reach the Earth, and it takes about five hours and 45 minutes to reach Pluto.

18. Frequent Flier

Frequent flier miles are a great way to get free flights and other benefits like priority bookings. **How much frequent-flier-mile money could Neil Armstrong have made on his trip to the Moon?** *(Since frequent flier miles are measured in miles, do this problem in English units.)*

ASK YOURSELF THIS...

A) What's the distance to the Moon?

B) How far did Neil Armstrong travel?

C) How much money do you earn per frequent-flier mile?

HELPFUL HINTS:

- From *Appendix C*, there are roughly 2.4×10^5 mi (~3.8 x 10 m) to the Moon. Since Armstrong made a round trip, his total distance traveled is twice that or about 4.8×10^5 mi.
- According to the Wikipedia entry for "frequent flier program," frequent flier miles are usually worth around $0.02 per mi.

CONSTRUCT A FORMULA:

Multiply Neil Armstrong's total distance traveled (4.8 x 10^5 mi) times the money earned per mile ($0.02 per mi) to get the total amount earned in frequent flier miles:

(# of miles traveled) x (money earned per mile)

MESSY MATH:

$$(4.8 \times 10^5 \text{ mi}) \times \left(\frac{\$0.02}{\text{mi}}\right) \approx \$9600$$

ANSWER:

Neil Armstrong could have gained about $9,600 worth of frequent-fliers miles on his trip to the Moon. According to Expedia, that's roughly worth 11 free round trips from New York to Hawaii, 10 free round trips from Boston to Tokyo, or less than one one-millionth the cost of a trip from Earth to the Moon on the Apollo program.

19. Sex Sells—So Who's Buying?

Of all of the things that fascinate us humans, sex seems to be at the top of the list. From dirty jokes to celebrity sex tapes, our curiosity is piqued whenever this cultural taboo is mentioned. **How many people are having sex right at this moment?**

ASK YOURSELF THIS...

A) How often does a person have sex?
B) How long does an intimate encounter typically last?
C) What percentage of the time are people having sex?
D) How many people are there?

HELPFUL HINTS:

- Averaging over a population that includes everything from priestly celibates to chronic nymphos, let's say that a typical person has sex about once every three weeks[x].
- The typical intimate encounter lasts about 15 minutes = 0.0015 weeks.
- From the previous two answers, it can be estimated that people

[x] I don't think any other problems in this book have caused me as much grief as the ones on sex and orgasms. While giving a talk on Fermi approximations, I was heckled by a woman who said my numbers in these problems were way too small. Perhaps she has a point—size definitely matters here—but I chose these numbers because roughly 50 percent of people told me they're way too small while the other 50 percent told me they're way too large. So if you find yourself agreeing with my heckler, then kudos on your awesome sex life, but you might not necessarily represent the mean.

have sex 15 minutes out of three weeks or about 0.05 percent of the time. Roughly the same percentage of people is having sex at any given time.

- There are 6.7×10^9 people in the world.

CONSTRUCT A FORMULA:

Multiply the fraction of people having sex (0.05% = 5.0×10^{-4} people having sex per person total) times the total number of people in the world (6.7×10^9 people) to get the number of people having sex:

(fraction of people having sex) x (# of people in the world)

MESSY MATH:

$$\left(5 \times 10^{-4} \frac{\text{people having sex}}{\text{person}}\right) \times (6.7 \times 10^9 \text{ people})$$

$$\approx 3.3 \times 10^6 \text{ people having sex}$$

ANSWER:

There are about 3.3 million people having sex at this very moment. (That's more than all the people in Chicago!)

20. Bad, Amélie! Go and Work on Your Arithmetic!

***Narrator**: Amélie still seeks solitude. She amuses herself with silly questions about the world below, such as "How many people are having an orgasm right now?"*

***Amélie**: Fifteen.*

—*Amélie in* Le Fabuleux destin d'Amélie Poulain

Let's see if we agree with Amélie. **How many people are having an orgasm this second?**

ASK YOURSELF THIS...

A) How often do people have orgasms?

B) How long does an orgasm last?

C) What percentage of our lives is spent having an orgasm?

D) How many people are in the world?

HELPFUL HINTS:

- The frequency of orgasms varies a great deal between men and women, between young and old, tops and bottoms, those who have an Internet connection and those who don't. Let's assume one orgasm per week.
- Like the frequency of orgasms, the length of time an orgasm lasts also varies considerably. Assume about 3 seconds or 5.0×10^{-6} weeks[xi].
- From the above two hints, we can calculate that 0.00050 percent of our lives is spent having orgasms. Roughly the same percentage of people is having an orgasm at any given moment.
- There are about 6.7×10^{9} people in the world.

CONSTRUCT A FORMULA:

Multiply the fraction of people having an orgasm (0.00050% = 5.0×10^{-6} people having an orgasm per person total) times the total number of people in the world (6.7×10^{9} people) to get the number of people having an orgasm:

(fraction of people having an orgasm) x (# of people in the world)

MESSY MATH:

$$\left(5 \times 10^{-6} \frac{\text{people having an orgasm}}{\text{person}}\right) \times (6.7 \times 10^{9} \text{ people})$$

$$\approx 3.3 \times 10^{4} \text{ people having an orgasm}$$

ANSWER:

We find that 3.3×10^{4} or 33,000 people are having an orgasm on Earth this very second. (Looks like Amélie forgot to carry the two!)

[xi] See the previous footnote.

21. Look Who's Talking

With all that sex people are having, you'd expect there to be lots of babies showing up. **How many babies are born each day?**

ASK YOURSELF THIS...

A) How many women are in the world?

B) How many children does a woman typically have?

C) How many days does a woman live?

D) If you pick a woman at random, what's the probability she's giving birth today?

HELPFUL HINTS:

- About half the world population, or 3.3×10^9 people, are women.
- In order to reproduce the population each generation, you need at least two or three children. Assume that the average woman has three children in her lifetime.
- A typical lifespan is about 80 years = 29,200 days.
- If women have three children during their lifetime, then on any given day the probability that a woman chosen at random is giving birth is 3 out of 29,200 or about 1/10,000[xii].

[xii] The reader may note that we haven't taken into account whether the woman chosen at random is of childbearing age. Since she is chosen at random, we have no way of knowing if she's 4 years old, 24 years old, or 79 years old. If we did take the age into account, the probability of giving birth would go down for the 4- and 79-year-olds and would go up for the 24-year-old, but the net result would be the same 1/10,000 number that we obtained above. Admittedly, this result may seem odd given our everyday experiences: while there's a 0.01 percent probability that a woman chosen at random will give birth today, there's a 100 percent probability that if you go up to a random woman and say, "There's a 0.01 percent chance you'll give birth today," she'll give you a disturbed look and back away slowly.

CONSTRUCT A FORMULA:

Multiply the ratio of births to women (1.0×10^{-4} births per woman) times the total number of women in the world (3.3×10^{9} women) to get the number of women giving birth today:

(ratio of births to women) x (# of women)

MESSY MATH:

$$\left(1.0 \times 10^{-4} \frac{\text{births}}{\text{women}}\right) \times (3.3 \times 10^{9} \text{ women}) \approx 3.3 \times 10^{5} \text{ births}$$

ANSWER:

There are about 330,000 births each day. Looking at the U.S. Census Bureau (www.census.gov/cgi-bin/ipc/pcwe) in 2009, we find the actual number to be very close at about 371,163 births per day.

22. Christmas's Secret War on Trees

There are those in the media who have claimed that secular progressives are leading an assault on Christmas. I'm not really sure what that means, but it seems that the biggest victims during the holiday season are trees—Christmas trees. **How much deforestation would result each year if people chopped down their trees from a forest rather than getting an artificial tree or getting one from a tree farm?**[xiii]

ASK YOURSELF THIS...

A) How many people get real Christmas trees each year?

B) How much area does a tree need to grow?

HELPFUL HINTS:

- There are 3.0×10^8 people in the United States. If only 10 % of people buy a Christmas tree every year, that's still 30 million trees.

[xiii] Some readers may question whether getting trees from farms is any better for the environment than getting them from a forest. According to a *Wired* magazine article cited in the Wikipedia entry for "tree farm":

> "A tree absorbs roughly 1,500 pounds of CO_2 in its first 55 years. . . . A well-managed tree farm acts like a factory for sucking CO_2 out of the atmosphere, so the most climate-friendly policy is to continually cut down trees and plant new ones. . . . Plant seedlings and harvest them as soon as their powers of carbon sequestration begin to flag, and use the wood to produce only high-quality durable goods like furniture and houses."

- Trees come in various sizes depending on their age. They can be up to a couple of meters wide and need some additional space to grow and get sunlight. Assume that in a forest, trees are spaced about 3 meters (~9.8 ft) apart. By squaring this distance, we see that each tree needs an area of about 10 m^2 to grow.

CONSTRUCT A FORMULA:

Multiply the number of trees cut down each year (3.0×10^7 trees) times the area per tree (10 m^2 per tree) to calculate the area of deforestation:

(# of trees) x (area per tree)

MESSY MATH:

$$(3.0 \times 10^7 \text{ trees}) \times \left(10 \frac{m^2}{\text{tree}}\right) \approx 3.0 \times 10^8 \, m^2$$

ANSWER:

This means that each year, we would cause about 300 km^2 (~120 sq mi) of deforestation.

23. The *Excellent* Code of Life!

In high school biology, we all learned that a sequence of DNA bases stores the code for building all living things. We're pretty complicated beings, so that code must be pretty big. But how big? **How long would all of the DNA in a human body be if you stretched it out?**

ASK YOURSELF THIS...

A) How many cells are in a human body?

B) How many DNA bases are in one cell?

C) What's the distance between one base and the adjacent base?

D) What's the total length of DNA in one cell?

HELPFUL HINTS:

- The number of cells varies a bit depending on how large you are, but a quick Web search shows that we have about 1.0×10^{13} cells in our body. (If you want to try calculating this on your own, use the fact that cells are several microns wide.)
- Every cell has one copy of our DNA tightly wound up inside it. Human DNA contains about 3.0×10^{9} base pairs.
- A quick Google search shows that consecutive base pairs in DNA are separated by 3.4×10^{-10} m (~1.1×10^{-10} ft). (Can you figure out a way to estimate that on your own?)
- Using the previous two hints, we can calculate that the total length of DNA in one cell is about 1.0 m.

CONSTRUCT A FORMULA:

Multiply the length of DNA per cell (1 m per cell) times the number of cells (1.0×10^{13} cells) to obtain the total length of DNA in your body:

(length of DNA per cell) x (# of cells)

MESSY MATH:

$$(1.0 \times 10^{13} \text{ cells}) \times \left(\frac{1.0 \text{ m}}{1 \text{ cell}}\right) \approx 1.0 \times 10^{13} \text{ m}$$

ANSWER:

This means there are 1.0×10^{13} m (~ 6 billion miles) of DNA in each of our bodies. That's long enough to wrap around the Earth 200,000 times or to reach the Sun about 70 times!

24. Mightier Than the Sword

The beginning of the school year can be a fun time. Ignoring the uneasiness that comes at the thought of another exam-filled nine months, there are the fresh new supplies of pens, bags, and Trapper Keepers that you get every September. But inevitably, the pens run out, get lost, or are stolen so that by the time June rolls around, you're left with a completely different gaggle of pens than you started with. What if you could maintain your original supply of pens? **How long of a line could you write before one pen ran out?**

ASK YOURSELF THIS...

A) If you don't lose it, how long does a pen last?

B) How much ink do you use per day?

HELPFUL HINTS:

- Assume that a six-pack of pens lasts about six months and that you lose half of these and use up the other half. It stands to reason that each pen lasts about two months = 60 days.
- If you take about 10 pages of notes per day with 20 lines per page and about 0.25 m (~10 in) of ink written per line, then you have used 50 m (~160 ft) of ink per day.

CONSTRUCT A FORMULA:

1) Multiply the length of ink used per day (50 m) times the number of days per month (30 days per month) to get the length of ink used per month.
2) Multiply this by the number of months per pen (two months) to calculate the length of ink used per pen:

(length of ink used per day) x (# of days per month) x (# of months per pen)

MESSY MATH:

$$\left(50\,\frac{\text{m}}{\text{day}}\right) \times \left(30\,\frac{\text{days}}{\text{month}}\right) \times \left(2\,\frac{\text{months}}{\text{pen}}\right) \approx 3000 \text{ m per pen}$$

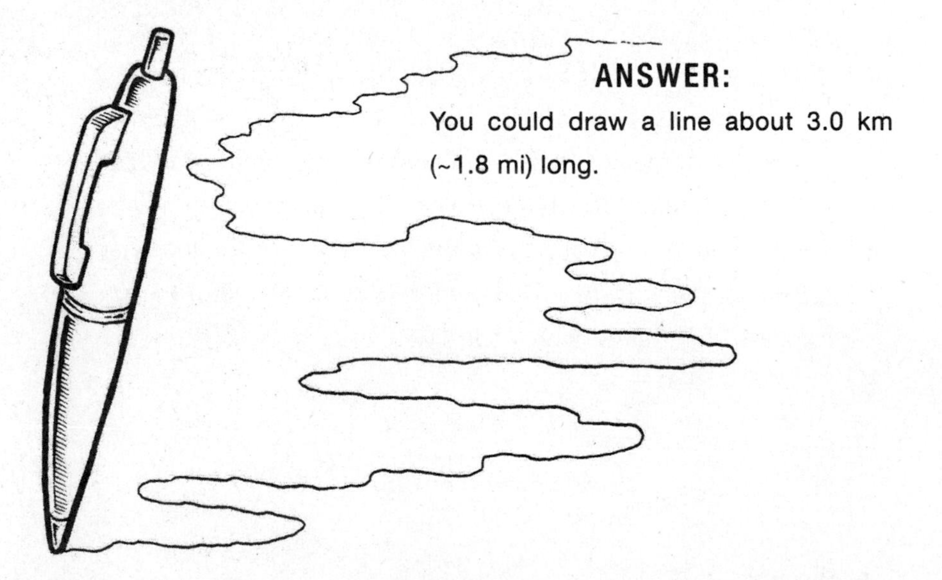

ANSWER:

You could draw a line about 3.0 km (~1.8 mi) long.

25. The Next Great Wall

While much of the talk in Washington today concerns sealing up the border between the United States and Mexico to deter illegal immigration, in the future we may be talking about sealing up the American-Chinese border. Due to continental drift, the Atlantic Ocean gets about 2 cm (~0.8 in) wider every year. Consequently, the Pacific Ocean gets 2 cm (~0.8 in) smaller every year. **How long will it be until the United States and China share a border?**

ASK YOURSELF THIS...

A) How wide is the Pacific?

B) How much does the Pacific shrink each year?

HELPFUL HINTS:

- Looking at a map, you can see that the Pacific Ocean is about 10,000 km = 1.0 x 10^9 cm (~6300 mi) wide.
- The Pacific shrinks about 2.0 cm (~0.78 in) per year.

CONSTRUCT A FORMULA:

Divide the width of the Pacific Ocean (1.0 x 10^9 cm) by the length shrunk per year (2.0 cm per year) to get the number of years until the United States borders China:

$$\frac{\text{(width of the Pacific)}}{\text{(length shrunk per year)}}$$

MESSY MATH:

$$\frac{(1.0 \times 10^{9}\ \text{cm})}{\left(2.0\,\frac{\text{cm}}{\text{yr}}\right)} \approx 5.0 \times 10^{8}\ \text{years}$$

ANSWER:

At a rate of 2 cm per year, it'll be only 500 million years before the United States borders China.

26. Wishing Well, O Wishing Well...

How many of us have tossed a penny into a well or fountain wishing for riches? A wishing well may not have the capacity for granting wishes, but does it have the capacity to hold a million dollars? **How many pennies could you fit in a well?**

ASK YOURSELF THIS...

A) What's the total volume of a wishing well?

B) What's the volume of a penny?

HELPFUL HINTS:

- Wells typically have a cross-sectional area that's on the order of 1 m^2 (~11 ft^2).
- Well depths vary a great deal depending on how deep down the water is found. Most wells for household use range anywhere from 30 to 300 m (~100–1000 ft). Assuming an intermediate well depth of 100 m, the total well volume is then about 100 m^3 (~3500 ft^3).
- You can calculate a penny's volume by multiplying its circular area times its height. The area of a circle is given by the formula

$A = \pi r^2$, where r is the radius and $\pi = 3.1415926...$ Using a radius of 1 cm and a thickness of 0.15 cm gives a volume of roughly $0.47\ cm^3 = 4.7 \times 10^{-7}\ m^3$ (~$0.028\ in^3$).

CONSTRUCT A FORMULA:

Divide the volume per well (100 m^3 per well) by the volume per penny (0.47 cm^3 per penny) to obtain the number of pennies per well:

$$\frac{\text{(volume per well)}}{\text{(volume per penny)}}$$

MESSY MATH:

$$\frac{\left(100\,\frac{m^3}{\text{well}}\right)}{\left(4.7 \times 10^{-7}\,\frac{m^3}{\text{penny}}\right)} \approx 2.1 \times 10^8 \text{ pennies per well}$$

ANSWER:

You could fit almost $2.1 million worth of pennies in the well.

27. Mr. Right

The self-help industry, loaded with love-finding tips, is a billion-dollar juggernaut. From seminars and books to online dating services to Julia Roberts movies, America spends compulsively on the relationship business, constantly searching for the special soulmate who may be only a Fung Wah bus ride away. I feel somewhat obligated (and a little ashamed) to tap into this vast money pool with this next calculation. **IF THERE IS ONLY ONE MR./MS. RIGHT FOR YOU, WHAT'S THE PROBABILITY OF MEETING HIM OR HER?**

ASK YOURSELF THIS...

A) How many people are in the world?

B) How many new people do you meet each day?

C) How many days will you live?

HELPFUL HINTS:

- There are about 6.7×10^9 people in the world.
- Assume you meet one new person every day.
- We have a lifespan of about 80 years. That's about 30,000 days and 30,000 people we'd meet.

CONSTRUCT A FORMULA:

Divide the number of people you'll meet (30,000 people) by the number of people in the world (6.7×10^9) to get the probability of meeting that special someone:

$$\frac{\text{(number of people you meet)}}{\text{(total number of people)}}$$

MESSY MATH:

$$\left(\frac{30{,}000 \text{ people}}{6.7 \times 10^9 \text{ people}}\right) \approx 4.5 \times 10^{-6} = 0.00045\%$$

ANSWER:

Assuming there are no mystical forces pulling you together, the probability of meeting special person X is then a finite, but small, 1 out of 200,000. It's no wonder many of us are spending to increase the odds!

28. Lincoln's Penny

The president is a very busy man. As leader of the free world, former President Bush had numerous responsibilities from balancing the budget to protecting our American citizens to spending 20 percent of his days in office vacationing at his Crawford ranch. One would imagine that he should be adequately compensated. If he were a king, he could collect his weight in gold, but perhaps instead we could just fill the Oval Office with pennies. **WHAT'S WORTH MORE MONEY: THE PRESIDENT'S ACTUAL SALARY OR THE NUMBER OF PENNIES THAT COULD FIT IN HIS OFFICE?**

ASK YOURSELF THIS...

A) What are the dimensions of the Oval Office?
B) What is the total volume of the Oval Office?
C) What is the volume of a penny?
D) What's the president's actual salary?

HELPFUL HINTS:

- The Oval Office measures 11 m (~35 ft 10 in) on the long axis, 8.8 m (~29 ft) on the short axis, and 5.6 m (~18 ft 6 in) in height.
- The total volume can be approximated from the dimensions above as 11 m x 8.8 m x 5.6 m ≈ 540 m^3 (~1.9 x 104 ft^3).
- In problem 26, we calculated the total volume of a penny to be about 4.7×10^{-7} m^3 (~0.028 in^3).
- A quick Google search shows several references listing the president's salary as $400,000 per year.

CONSTRUCT A FORMULA:

Divide the volume of the Oval Office (540 m^3) by the volume per penny (4.7×10^{-7} m^3) to obtain the number of pennies:

$$\frac{\text{(volume of the Oval Office)}}{\text{(volume of a penny)}}$$

MESSY MATH:

$$\frac{\left(540\ \text{m}^3\right)}{\left(4.7 \times 10^{-7}\ \frac{\text{m}^3}{\text{penny}}\right)} \approx 1.1 \text{ billion pennies in the Oval Office}$$

ANSWER:

About $11 million worth of pennies can fit in the Oval Office. That's 30 times what the President makes.

29. The Wonder of Wrapping Paper

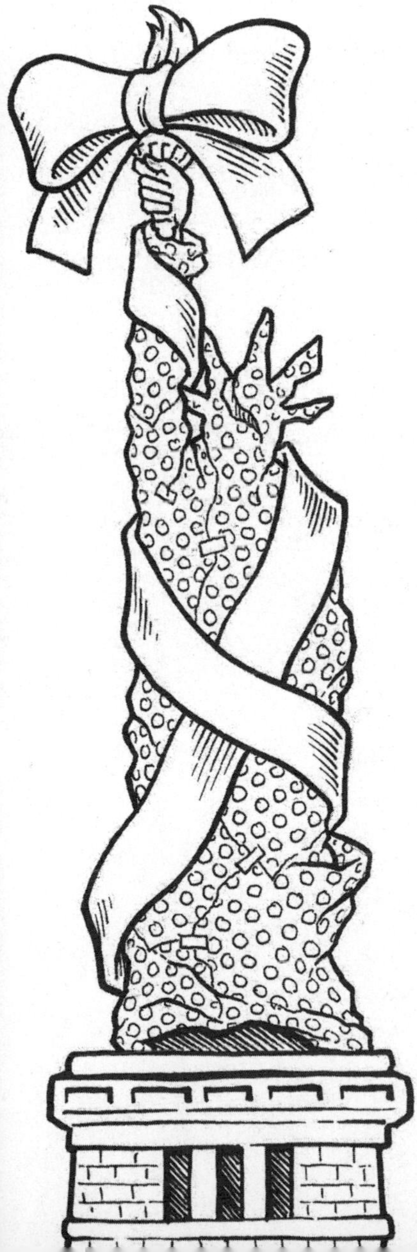

Christo and Jeanne-Claude have made a name for themselves with their elaborate environmental installation art. Their works include wrapping the Reichstag in Berlin and the Pont Neuf Bridge in Paris with fabric. This obsession with wrapping is curious. Instead of constructing *The Gates* in Central Park, they could have wrapped the Statue of Liberty. And they could have used wrapping paper to save money. **How much would it cost to wrap the Statue of Liberty in wrapping paper?**

ASK YOURSELF THIS...

A) How tall is the Statue of Liberty?

B) How wide and thick is she?

C) What total area needs to be wrapped?

D) How much does wrapping paper cost?

HELPFUL HINTS:

- Lady Liberty's about the height of a 10-story building. According to Wikipedia, her height from the ground to the tip of the torch is 46 m (~150 ft).
- She's about 10 m (~33 ft) wide and 10 m (~33 ft) thick.
- Since the area of her front is about equal to the area of her back and sides, the total area that needs to be wrapped will be about four times the area of a side or about 4 x 460 m^2 = 1,800 m^2 (~19,000 sq ft).
- If you purchase in bulk, you can buy a 300 m x 1 m roll of white wrapping paper for as little as \$60 per roll. This means the cost per unit area is \$0.20 per m^2.

CONSTRUCT A FORMULA:

Multiply the total area to be wrapped (1,800 m^2) by the cost per unit area of wrapping paper (\$0.20 per m^2) to obtain the total price:

(total area to be covered) x (cost of wrapping paper per unit area)

MESSY MATH:

$$(1800\ \text{m}^2) \times \left(\frac{\$0.20}{\text{m}^2}\right) \approx \$360$$

ANSWER:

Using wrapping paper instead of fabric, it would cost Christo and Jeanne-Claude only about \$360 to wrap Old Lady Liberty, not too much (especially given the cost of their other projects).

30. Monkeys at Typewriters

During the 1800s, the "Great Debate" took place between Anglican Archbishop Samuel Wilberforce and evolutionist-agnostic Thomas Huxley. Wilberforce defended his belief in a creator by asserting that the design we see in nature requires a designer. Huxley countered with the point that given sufficient time, all possible combinations of matter, including those necessary to produce a man, would eventually occur by random chance, just as a monkey given an infinite amount of time typing at a typewriter would—by random chance—eventually reproduce the works of Shakespeare. **How long would it take one monkey randomly typing at a typewriter to reproduce the entire works of Shakespeare?**

ASK YOURSELF THIS...

A) What's the probability of hitting the correct key?

B) What's the probability of hitting all the correct keys consecutively?

C) How many keys do you need to hit to type the complete works of Shakespeare?

D) How much time does it take to write all of Shakespeare's plays?

HELPFUL HINTS:

- A typewriter has roughly 40 keys (all of the letters plus periods, spaces, semicolons, etc.). If, for simplicity, we do not consider capitalization, then the probability of hitting the correct key is 1 out of 40 or 0.025 = 2.5%.
- The probability of hitting one correct key is 1/40, the probability of hitting two correct keys is $(1/40)^2$, the probability of hitting three correct keys is . . . . the probability of hitting *n* correct keys is $(1/40)^n$.
- In a book of Shakespeare, there are about 40 characters (as in letters and punctuation) per line, 30 lines per page, 80 pages per play, and 37 plays. Multiplying all these together, we get a total of about 3.6×10^6 characters in the complete works of Shakespeare.
- Assume that a monkey can type a letter every 0.1 seconds and that he never tires or stops. The time to write all of Shakespeare's works can be found by multiplying the total number of characters found above (3.6×10^6 characters) times the time spent on each character (0.1 s per character). From this, we obtain a total of 3.6×10^5 s = 4.1 days for a monkey to type the complete works of Shakespeare.

CONSTRUCT A FORMULA:

1) To obtain the probability of correctly typing out the complete works of Shakespeare, raise the probability of hitting one correct key (0.025) to a power given by the number of characters (3.6×10^6 characters).

2) You can then approximate the number of days it would take to correctly type the works of Shakespeare by dividing the number of days it takes to type out all the characters you need (4.1 days) by the probability of typing all the correct keys:

$$\frac{\text{(time it takes to write all of Shakespeare)}}{\text{(prob. of hitting the right key)}^{\text{(\# of keys you need to hit)}}}$$

MESSY MATH:

$$\frac{(4.1 \text{ days})}{(0.025)^{3.6 \times 10^6}} \approx 3.8 \times 10^{5,767,416} \text{ days}$$

ANSWER:

It would take about $3.8 \times 10^{5,767,416}$ days. This number is so large that were you to attempt to write it out, it alone would be about a 5,000-page book!

31. Ghost Gobblers

In 1983, Bill Murray, Dan Aykroyd, and Harold Ramis "spooked" audiences across America with the instant classic *Ghostbusters*. One of the most memorable moments of the movie comes in the climactic battle between our three heroes and an immense marshmallow monster. Yes, audiences loved Mr. Stay Puft, with his round cherubic face and boyish sailor outfit. And yet the question begs to be asked: **How long would it take to eat him?**

ASK YOURSELF THIS...

A) How tall is Mr. Stay Puft? How wide is he? How thick?

B) What's his total volume?

C) How many Calories are in a marshmallow?

D) What's the volume of a marshmallow?

E) How many Calories do we eat in a day?

HELPFUL HINTS:

- In the movie Mr. Stay Puft is about the size of a 10-story building. According to a Boston architecture firm, each story of a typical 10-story building is about 4 m (~13 ft) high.
- A typical city building might be about 20 m (~66 ft) wide. The thickness is about the same as the width.
- Assume his total volume is just his height times his width times how thick he is.
- Kraft marshmallows are 25 Calories each.
- A 3 cm (1.2 in) tall cylindrical marshmallow with a 1 cm (~0.4 in) radius has a volume of roughly 10 cm^3 or $1.0x10^{-5}$ m^3 (~0.61 In^3).
- The Food and Drug Administration recommends 2,000 Calories per day as a healthy average.

CONSTRUCT A FORMULA:

1) Multiply the height per story (4 m/ story) by the number of stories (10 stories) to find that Mr. Stay Puft's height is about 40 m (~130 ft).

2) Next, multiply this height times his width (20 m) and times his thickness (20 m) to get his volume $1.6 \times 10^3 m^3$ (~$5.7 \times 10^4 ft^3$).

3) Divide the number of Calories (25 Cal) by the volume of a marshmallow (1.0×10^{-5} m^3) to obtain the number of Calories per unit volume of a marshmallow to be about 2.5×10^6 Cal/m^3.

4) You can then easily calculate the number of days it would take to eat him by multiplying Mr. Stay Puft's volume by the number of Calories per unit volume in a marshmallow and then dividing by the number of Calories a person eats per day (2,000 Cal/ day):

$$\frac{\textbf{(vol. of Mr. SP) x (\# Cal. per unit vol.)}}{\textbf{(\# days per Cal.)}}$$

MESSY MATH:

$$\frac{\left(1.6 \times 10^{4}\ \text{m}^{3}\right) \times \left(2.5 \times 10^{6}\ \frac{\text{Cal}}{\text{m}^{3}}\right)}{\left(2000\ \frac{\text{Cal}}{\text{day}}\right)} \approx \text{20 million days or 55,000 years}$$

ANSWER:

Without a positron collider, it would take one person 55,000 years or 55,000 people one year to finish him off.

32. Eat It, Spider-Man!

When it comes to superhero movies, one must always maintain some suspension of disbelief. For instance, in the *Spider-Man* series, it's perfectly natural for audiences to believe that getting bitten by a radioactive spider gives you super powers, but are we really expected to believe all of that web came out of one body? Isn't there some physical law that says matter can't be created on the fly? **How many pounds of food does Peter Parker have to eat to produce an equivalent amount of web each day?**

ASK YOURSELF THIS...

A) How long is each web that gets shot?

B) How many webs does Spider-Man shoot each day?

C) How thick is each web Spidey shoots?

D) What's the density of spider-silk?

HELPFUL HINTS:

- In the movies it appears that webs can be about 20 m = 2.0 x 10^3 cm (~66 ft) long.
- If each web takes him about 20 m (~66 ft), then he needs to shoot about 80 webs to travel a mile, which is a reasonable distance for him to travel in a day.
- In the movies Spider-Man's webs appear to be about 1.0 cm (~0.39 in).[xiv]
- A quick web search finds that the density of spider silk is about 1.3 g/ cm^3 (~0.58 oz/ in^3).

[xiv] The movie does not use real spider silk. Real spider silk has a tensile strength of about 1.3 GPa, which is almost the same as high-grade steel. However, since it is less dense than steel, it need not be as thick as the web depicted in the movies. To support a 180-lb person, the web would need to be only about one millimeter thick.

CONSTRUCT A FORMULA:

1) Square the web thickness (1.0 cm) and multiply by the length per web shot (2.0×10^3 cm per webs shot) to get the total volume of each web shot.
2) Multiply by the density of spider silk (1.3 g/ cm^3) to obtain the mass of each web shot.
3) Finally, multiply by the total number of webs shot (88 webs shot) to obtain the total mass of webs shot per day:

(# of webs shot) x (web thickness)^2 x (web length) x (density of spiderwebs)

MESSY MATH:

$$(\text{80 webs}) \times (1.0\text{ cm})^2 \times \left(2000\,\frac{\text{cm}}{\text{web}}\right) \times \left(1.3\,\frac{\text{g}}{\text{cm}^3}\right) \approx 210{,}000\text{ g}$$

ANSWER:

Each day, he would need to eat 210 kg or about 460 lbs just to have enough material to make all that web.

33. Environmentally Speaking

In his 2006 State of the Union address, former President Bush said, "America is addicted to oil, which is often imported from unstable parts of the world."

This addiction would seem to be the root of oil companies' record profit, global warming, and America's strategic interest in the Middle East. One way to eliminate these problems would be to change our status from energy consumer to energy provider. Some have argued that America needs a solar energy research program the size of the Manhattan Project. Before this option can face opposition from lobbyists within the oil industry, one should really consider how feasible a proposition it is from a physical standpoint. **How much area would you need to have solar panels powering the entire United States?**

ASK YOURSELF THIS...

A) How much power does America use?

B) How much power from sunlight falls on a square meter of Earth's surface?

C) How efficient are solar cells?

HELPFUL HINTS:

- According to *hypertextbook.com*, Americans use energy at a rate of about 3.2×10^{12} W.[xv]
- If you haven't taken a physics class, you'll likely have a difficult time approximating the amount of power sunlight has. It may even be difficult to search for it on the Web if you don't know which words to look for. When it comes to solar energy, the phrase you want to remember is "solar flux," which is a measure of the amount of energy that comes per unit area per unit time. A quick Google search shows that the solar flux at the surface of the Earth during the daytime is 1,300 W/ m^2. Since it's night half the time, the average power will be closer to 650 W/ m^2.
- In 2007 a University of Delaware research team set a record with 42.8 percent efficiency.

CONSTRUCT A FORMULA:

1) To obtain the total power you get from the solar panels, you must multiply the efficiency (0.428) by the power per unit area of sunlight falling to the Earth (650 W/ m^2).

2) You can then calculate the total area you need to power the United States by dividing this into the U.S. power consumption (3.2×10^{12} W):

$$\frac{\text{(U.S. power consumption)}}{\text{(efficiency)} \times \text{(power per unit area)}}$$

[xv] The "W" stands for Watt (after Scottish inventor James Watt.) It's a unit of power that is most often associated with light bulbs. One Watt is equivalent to one kg m^2 s^{-3}.

MESSY MATH:

$$\frac{(3.2 \times 10^{12}\ \mathrm{W})}{(0.428) \times \left(650\ \frac{\mathrm{W}}{\mathrm{m}^2}\right)} \approx 1.2 \times 10^{10}\ \mathrm{m}^2$$

ANSWER:

You'd need about 12,000 km^2 (~4,600 sq mi) or a solar panel that's almost three times the size of Rhode Island.

By the way, only a small percentage of the Sun's radiated energy actually hits the Earth. Absorbing only one second of the total power emitted by the Sun would be enough energy to power the United States for more than 1 million years.

34. Location Is Everything

One would imagine that the good people of Rhode Island wouldn't want to sacrifice their state just so we can build a giant solar panel there, so it behooves us to find another place for the panels. According to Scott Brushaw on *solarroadways.com*, if you could build sturdy enough panels, you could make roads out of solar panels. **If we made highways out of solar panels, would they produce enough energy to power the United States?**

ASK YOURSELF THIS...

A) How many miles of highways are in the United States?

B) How wide is a typical highway?

C) How much total area do you need to power the United States with solar panels?

HELPFUL HINTS:

- According to Wikipedia, the Interstate Highway System has a total length of more than 75,000 km (~46,000 mi).
- Roads are about 33 m (~110 ft) wide.
- From the previous problem, you need at least 1.2×10^{10} m^2 to power the United States.

CONSTRUCT A FORMULA:

Multiply the width of the road (33 m) by the length of all the roads (7.5 x 10^7 m) to get the total area of all the roads:

(width of a road) x (length of all the roads)

MESSY MATH:

$$(33 \text{ m}) \times (7.5 \times 10^7 \text{ m}) \approx 2.5 \times 10^9 \text{ m}^2$$

ANSWER:

We'd get about one-fifth of the energy we need if we tiled all the highways with solar panels, but if we tiled the rest of the paved roads, we could easily come up with the rest of the energy we need.

35. It's an Investment

If the proposition of a completely solar-powered North America seems too optimistic to be true, there is the small matter of price. **How much would the solar-powered system described above cost taxpayers?**

ASK YOURSELF THIS...

A) How much power does the United States consume?

B) How much does a solar panel cost to install per watt?

HELPFUL HINTS:

- From example 33, Americans use energy at a rate of about 3.2×10^{12} W.
- There are various web pages that list the cost of photovoltaic panels. Depending on the type, they generally cost somewhere around $6.50 per W.

CONSTRUCT A FORMULA:

Multiply the cost per watt ($6.50 per W) times the number of watts consumed (3.2×10^{12} W) to obtain the total cost:

(cost per watt) x (# of watts consumed)

MESSY MATH:

$$\left(\frac{\$6.50}{W}\right) \times (3.2 \times 10^{12}\ W) \approx \$2.0 \times 10^{13}$$

ANSWER:

The total cost of all the panels we need to support our present energy expenditure would be $20 trillion. (At the time of this writing, this is seven times larger than the national budget and almost twice the national debt!)

36. How Many Licks Does It Take?

It's amazing what bored college students will do on a Friday night during finals when there are no parties going on. Recently in Cambridge, Massachusetts, a group of students from a small liberal arts college, or "Harvard," set out on a quest: to find out how many licks it takes to get to the Tootsie Roll center of a Tootsie Pop. Their answer: 900. But could those Cantebrigians have used a smarter way to figure out the answer? **How many licks does it take to get to the Tootsie Roll center of a Tootsie Pop?**

ASK YOURSELF THIS...

A) What happens to a Tootsie Pop when it's licked?

B) How much do we lick off at once?

C) What's the smallest thing the naked human eye can see?

D) How thick is a Tootsie Pop?

HELPFUL HINTS:

- If a Tootsie Pop has only been licked once, you can just barely tell it's been licked. That being the case, the amount of material removed is probably about the size of the smallest thing we can see.
- The smallest objects that human eyes can see are about 10 microns = 1.0×10^{-3} cm (~0.0004 in) in size. This is about the size of a drop of mist or a speck of dust.
- A Tootsie Pop is about 2.5 cm (~1.0 in) thick. The distance to the Tootsie Roll center is about one-third this or about 0.8 cm.

CONSTRUCT A FORMULA:

Divide the distance to the center of the Tootsie Pop (0.8 cm) by the length of material removed per lick (1.0×10^{-3} cm per lick) to calculate the number of licks needed to get to the center of a Tootsie Pop:

$$\frac{\text{(distance to the center of a Tootsie Pop)}}{\text{(\# of cm removed per lick)}}$$

MESSY MATH:

$$\frac{(0.8\ \text{cm})}{\left(1.0 \times 10^{-3}\,\frac{\text{cm}}{\text{lick}}\right)} \approx 800\ \text{licks}$$

ANSWER:

It takes about 800 licks. Those Harvard students didn't do too badly.

37. Steve Martin's Wish

In a bit on *Saturday Night Live,* Steve Martin declared that if he could have one wish during the holiday season, it would be that all the children of the world join hands and sing together in the spirit of harmony and peace. (However, upon realizing that the logistics of the matter would be quite difficult to implement, he decided that he should instead wish for a 31-day orgasm, $30 million a month to be given to him tax-free in a Swiss bank account, all-encompassing power over every living being in the entire universe, and revenge against his enemies so they could die like pigs.) How impractical would it be to get all the children of the world to hold hands and sing together in the spirit of peace and harmony? **SPECIFICALLY, HOW MANY TIMES COULD ALL THE CHILDREN ENCIRCLE THE GLOBE IF THEY HELD HANDS IN A CONTINUOUS LINE?**

ASK YOURSELF THIS...

A) How many people are in the world?
B) How many children are in the world?
C) How far can one child stretch from hand-to-hand?
D) What's the radius of the Earth?
E) What's the circumference of the Earth?

HELPFUL HINTS:

- There are roughly 6.7×10^9 people in the world.
- About half the population or roughly 3.3×10^9 people are children.
- Children come in all sizes. Assume there's a length of 1.2 m (~4 ft) from hand to hand in a child stretching out his or her arms.

- From *Appendix C,* the radius of the Earth is 6,378 km (~3,963 mi) or about 6.4×10^6 m.
- The circumference of a circle is 2π times the radius. Using the radius of the Earth, the circumference of the Earth can be calculated to be 4.0×10^7 m (~25,000 mi).

CONSTRUCT A FORMULA:

1) Multiply the number of children (3.3×10^9 children) times the length per child (1.2 m per child) to obtain the total length of children.
2) Divide this by the circumference of the Earth (4.0×10^7 m) to obtain the total number of times children holding hands can circle the Earth:

$$\frac{\text{(\# of children) x (\# of meters per child)}}{\text{(circumference of Earth)}}$$

MESSY MATH:

$$\frac{(3.3 \times 10^9 \text{ children}) \times \left(1.2 \dfrac{\text{m}}{\text{child}}\right)}{(4.0 \times 10^7 \text{ m})} \approx 100 \text{ times}$$

ANSWER:

All the world's children holding hands could loop around the Earth 100 times.

38. *Mmm . . .* Mouth-Watering Savior

When I was little and in Catholic school, I learned about Holy Communion. The little wafer and wine we would eat were supposed to be the body and blood of Jesus. At the time, I thought this was a little odd for a few reasons. One was that I remembered reading somewhere that cannibalism was a sin, and I figured it would be far worse if you cannibalized Jesus. Second, I always wondered what part I was getting, and I was really hoping it wasn't a toe. Lastly, I wondered with all the Catholics in the world, shouldn't Jesus' body have run out by now? Let's satisfy my inner first grader's curiosity. **How many Masses would it take until we ran out of all the Eucharist wafers you could make out of one Jesus?**

ASK YOURSELF THIS...

A) How much does Jesus weigh?

B) How much does a Eucharistic wafer weigh?

C) How many people can fit in a church?

HELPFUL HINTS:

- People were smaller back then and Jesus fasted a lot. A fairly svelte Jesus weighs about 55 kg (~120 lbs).
- Each Eucharistic wafer weighs about 1.0 g = 0.001 kg (~0.035 oz).
- An extremely large church, say the church of St. Pius X in Lourdes, France, can accommodate 20,000 members. A more modest church fits around 100 people in a Mass. Assume each member gets one Eucharistic wafer.

CONSTRUCT A FORMULA:

1) Divide the weight of Jesus (55 kg) by the weight per wafer (0.001 kg) to obtain the number of wafers one Jesus can make.
2) Divide this by the number of wafers eaten per Mass for a large (20,000 wafers per Mass) and small (100 wafers per Mass) church to obtain the number of Masses that can be made from one Jesus:

$$\frac{\text{(weight of Jesus)}}{\text{(weight per wafer)} \times \text{(\# of wafers per Mass)}}$$

MESSY MATH:

For a large Mass:

$$\frac{(55\ \text{kg})}{\left(0.001\ \frac{\text{kg}}{\text{wafer}}\right) \times \left(20{,}000\ \frac{\text{wafers}}{\text{Mass}}\right)} \approx 3 \text{ large Masses}$$

or

For a small Mass:

$$\frac{(55\ \text{kg})}{\left(0.001\ \frac{\text{kg}}{\text{wafer}}\right) \times \left(100\ \frac{\text{wafers}}{\text{Mass}}\right)} \approx 550 \text{ small Masses}$$

ANSWER:

Jesus would last three Masses at the Church of St Pius X, or almost 10 years of Masses at a small church.

39. The Nailanator

"Conservative" Comedy Central pundit Stephen Colbert has quite a list of accolades: Peabody Awards, Emmys, his own Ben & Jerry's Ice Cream Flavor. Since 2005, Colbert has made a career out of "nailing" guests on his hit show *The Colbert Report.* Whether it's getting Congressman Robert Wexler to confess to the joys of cocaine use or using the sheer attractive powers of his face to pilfer billionaire philanthropist Richard Branson's drinking water, there seems to be no limit on just how far Colbert will go to bring down the hammer. **If he maintains his torrid pace, how many guests will Colbert nail in his career?**

ASK YOURSELF THIS...

A) How many years will *The Colbert Report* be on?

B) How many weeks does the show run each year?

C) How many guests does Colbert nail each week?

HELPFUL HINTS:

- Since Colbert's in his mid-40s, *The Report* could easily be on another 20 years.
- *The Report* is on about 40 weeks per year.
- There are four episodes per week. Usually the show has one guest, sometimes two, meaning about five people are nailed each week.

CONSTRUCT A FORMULA:

Multiply the number of years *The Colbert Report* will be on (20 yrs) times the number of weeks per year (40 weeks per yr) times the number of nails per week (5 nails per wk) to obtain the total number of nails Colbert will have in his career:

(# of years) x (# of weeks per year) x (# of nails per week)

MESSY MATH:

$$(20\text{ yrs}) \times \left(40\ \frac{\text{wks}}{\text{yr}}\right) \times \left(5\ \frac{\text{nails}}{\text{wk}}\right) \approx 4{,}000\text{ nails}$$

ANSWER:

By 2029, Colbert should have shattered Bob Vila's record for most nails in a career by a television show host.

40. Homer the Great

The Simpsons, considered by many to be the greatest sitcom of all time, has been entertaining audiences since 1989. After 23 Emmys, 26 Anny Awards, and one Peabody, well over 400 episodes have been produced. That's a lot of animation. **How far would all the film frames that make up the 400-plus episodes of *The Simpsons* stretch?**

ASK YOURSELF THIS...

A) How many episodes have aired?

B) How many minutes does an episode last?

C) How many frames are drawn for each second of animation?

D) How large is each frame?

HELPFUL HINTS:

- Since *The Simpsons* first aired in 1989, there have been about 450 episodes.
- Each episode contains about 20 minutes (~1,200 s) of footage once you take out the commercials.
- Film is projected at 24 frames per second, but most animation studios don't do that many drawings. Generally, 12 drawings per second are used in animation.
- Film gauges typically come in 8 mm, 16 mm, and 35 mm. Assume *The Simpsons* uses a 16 mm = 0.016 m (~0.63 in) film reel.

CONSTRUCT A FORMULA:

1) Multiply the number of episodes (450 episodes) times the number of seconds per episode (1,200 s per episode) to obtain the length of time each episode lasts.
2) Multiply this by the number of frames per second (12 frames per s) to obtain the total number of frames used in all episodes.
3) Finally, multiply the number of frames by the length per frame (16 mm per frame) to obtain the total length of all the frames:

(# episodes) x (# seconds per episode)
x (# frames per second) x (length per frame)

MESSY MATH:

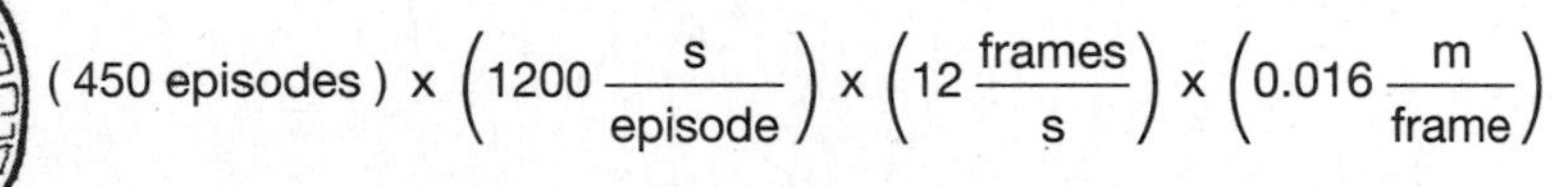

$$(\text{450 episodes}) \times \left(1200 \frac{\text{s}}{\text{episode}}\right) \times \left(12 \frac{\text{frames}}{\text{s}}\right) \times \left(0.016 \frac{\text{m}}{\text{frame}}\right)$$

$$\approx 100{,}000 \text{ m}$$

ANSWER:

There are about 100 km (~ 62 miles) of film slides, enough to run across Rhode Island.

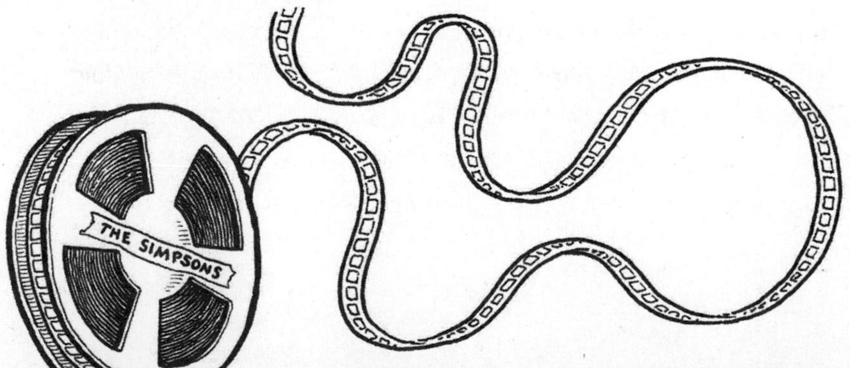

41. The Boston Tea Party

My childhood recollections of history class are somewhat scant. I vaguely remember something about the good people of Boston throwing the Brits' tea into the harbor because they didn't like tea. (At least I think it was tea. It was either tea or taxes, but I'm pretty sure it was tea.) If my memories of this particular history class are less than perfect, then the images I have of this event are downright dreadful. First, I imagined the tea not as hot tea, but as Lipton Iced Tea, which I'm pretty sure they didn't have back then. Second, I thought it would have made the whole harbor taste like salty iced tea. Would there even have been enough tea to taste it? **How many grams of tea would you need to make the Boston Harbor palatable?**

ASK YOURSELF THIS...

A) What's the volume of Boston Harbor?

B) What's the volume of a cup of tea?

C) What's the mass of a tea bag?

HELPFUL HINTS:

- By looking at a map, you can estimate the harbor to have a total area of about 200 km^2 = 2.0×10^{12} cm^2 (~510 mi^2). The depth of the harbor varies, but on average should be about 15 m = 1,500 cm (~49 ft). This gives a total volume of 3.0×10^{15} cm^3 (~1.1×10^{11} ft^3).
- A 10-oz cup of tea has a volume of about 300 cm^3 (~18 in^3).
- Each tea bag weighs roughly 1.5 g (~0.053 oz).

CONSTRUCT A FORMULA:

1) Divide the weight per tea bag (1.5 g per tea bag) by the volume of water used made per bag (300 cm^3 per tea bag) to obtain weight of tea bags used per unit volume of water.
2) Multiply this by the volume of the harbor (3.0×10^{15} cm^3) to obtain the total weight of tea bags needed to fill Boston Harbor:

$$\frac{\text{(weight per tea bag)} \times \text{(volume of the harbor)}}{\text{(volume of water used per tea bag)}}$$

MESSY MATH:

$$\frac{\left(1.5\,\frac{g}{\text{teabag}}\right) \times (3.0 \times 10^{15}\ \text{cm}^3)}{\left(300\,\frac{\text{cm}^3}{\text{teabag}}\right)} \approx 1.5 \times 10^{13}\ \text{g}$$

ANSWER:

You need 1.5×10^{10} kg or more than 10 million tons of tea to make the Boston Harbor even remotely palatable.

According to Wikipedia, the entire world produced only 3.15 million tons of tea in 2003.

42. Chess and Rice

In eighth grade my math teacher, Mr. Larkin, told us a story about the invention of chess. A long time ago in a faraway land, the king was pretty bored because back then they didn't have iPods, flat screens TVs, or *How Many Licks?* books to keep them entertained. One day, a servant of the king invented a new game that he called "chess." The king was so enamored with the game that he told the servant he could have anything he wanted. All the servant asked for was that for the first square on the chess board he would get two grains of rice, and for the second square he would get four grains of rice, and on the third he would get eight grains of rice, and so on so that each square would get double the amount of rice as the previous one until all 64 squares were covered. The king granted it immediately, but then changed his mind and promptly had the servant executed by decapitation for being greedy and gluttonous. Why? **WHAT VOLUME OF RICE WOULD HAVE BEEN ON JUST THE LAST SQUARE?**

ASK YOURSELF THIS...

A) How many grains of rice are on each square?

B) What's the volume of a grain of rice?

HELPFUL HINTS:

- The first square has 2^1, the second has 2^2 the third has 2^3 . . . the 64th has 2^{64} grains of rice.
- Experience tells us that a grain of rice has dimensions of about 1 mm x 1 mm x 1 cm, which results in a total volume of 10 mm^3 = 1.0×10^{-8} m^3 (~6.1×10^{-4} in^3).

CONSTRUCT A FORMULA:

Multiply the number of grains needed on the last square ($2^{64} = 1.8 \times 10^{19}$ grains) times the volume per grain (1.0×10^{-8} m^3 per grain) to obtain the total volume of rice on the last square:

(# grains needed on the last square) x (volume per grain)

MESSY MATH:

$$(1.8 \times 10^{19} \text{ grains}) \times \left(1.0 \times 10^{-8} \frac{m^3}{\text{grain}}\right) \approx 1.8 \times 10^{11}\ m^3$$

ANSWER:

You'd need 1.8×10^{11} m^3 grains of rice. If we laid them all flat on the ground, there would be more than enough rice grains to cover the dry land on Earth, and that's just from the last square!

43. Google, Googol, Googolplex

U.S. Senator and pseudo-cyberspace philosopher Ted Stevens is famous for noting of the Internet, "It's not a truck. It's a series of tubes." **If you were to convert the websites that travel through these "tubes" into an encyclopedia, how tall would it be?**

ASK YOURSELF THIS...

A) How many web pages are there?

B) How thick is a page of paper?

C) How many pages do you need to print a website?

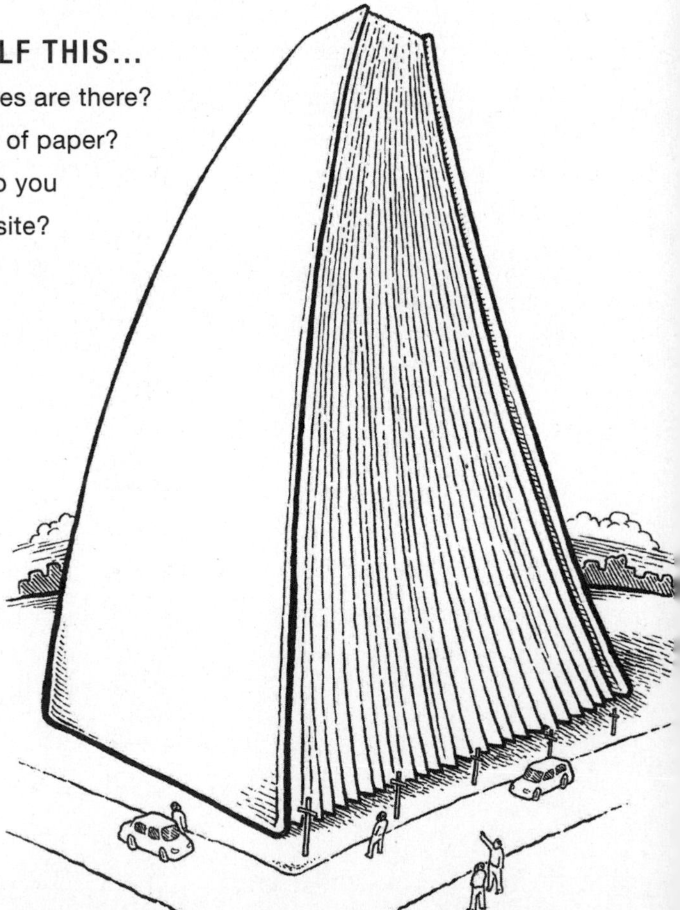

HELPFUL HINTS:

- According to Wikipedia, there were at least 100 million websites as of March 2008. (Challenge: Can you find a clever way to come up with that figure without looking it up?)
- A 100-page book is about 1.0 cm (0.39 in) thick, so one page is about 0.01 cm thick.
- Web pages vary considerably in size, but most would certainly be less than 100 pages and more than 1 page if you printed them out. Assume that, on average, there are 10 printed pages per web page.

CONSTRUCT A FORMULA:

Multiply the number of websites (1.0×10^8 websites) times the number of pages per site (10 pages per site) times the width per page (0.01 cm per page) to calculate the total width of a book containing the content of all websites:

(# of websites) x (# of pages per website) x (length per page)

MESSY MATH:

$$(1.0 \times 10^8 \text{ websites}) \times \left(10 \frac{\text{pages}}{\text{website}}\right) \times \left(0.01 \frac{\text{cm}}{\text{page}}\right) \approx 1.0 \times 10^7 \text{ cm}$$

ANSWER:

The book would have to be about 100 km (~62 mi) long. It would take about one hour to drive from one end to the other.

44. The Great Seattle Umbrella Rig

Seattle has a reputation for being "the rainy city."[xvi] In 1953, it set a record with 33 consecutive days of rain. The rain can't be all that bad, however, since Seattle is also one of America's most intelligent cities, having the highest percentage of college graduates of any major U.S. city. The mathematical minds of these intellectual inhabitants may wonder why they don't just build a rig of umbrellas above the city. **How much would it cost to cover Seattle with a rig of umbrellas?**

ASK YOURSELF THIS...

A) What's the total area of Seattle?

B) How much area does an umbrella take up?

[xvi] Despite its reputation, Seattle is not even listed as one of the top 10 rainy cities in the United States.

HELPFUL HINTS:

- You can estimate the total area by looking at a map of Seattle. Seattle has an area of roughly $2.2 \times 10^8\ m^2$ (~84 sq mi).
- Umbrellas cover an area of about $1.0\ m^2$ (~11 ft^2), and you can find them at a dollar store.

CONSTRUCT A FORMULA:

1) Divide the total area of Seattle ($2.2 \times 10^8\ m^2$) by the area per umbrella ($1.0 m^2$ per umbrella) to get the number of umbrellas it takes to cover all of Seattle.
2) Multiply this by the cost per umbrella ($1 per umbrella) to get the total cost of building an umbrella rig:

$$\frac{\text{(total area of Seattle) x (cost per umbrella)}}{\text{(area per umbrella)}}$$

MESSY MATH:

$$\frac{(2.2 \times 10^8\ m^2) \times \left(\frac{\$1}{\text{umbrella}}\right)}{\left(1 \frac{m^2}{\text{umbrella}}\right)} \approx \$2.2 \times 10^8$$

ANSWER:

That gives about 220 billion umbrellas for $220 billion (about half the military budget of the United States), not to mention construction fees and operation costs. Sorry, Seattle...no umbrella for you!

45. Insignificance

"I'm not sure, but he seems to be inordinately fond of beetles."

—J.B.S. Haldane, when asked what the study of science taught him about "The Creator."

Haldane's comment reflects the fact that there are more than 350,000 known species of beetle, making up 40% of all insects and roughly six times the number of all known vertebrate species. Humans typically think of themselves as the dominant creature on the planet, but if there is a creator, he might very well enjoy beetles far more than man. He certainly didn't make very many of us. **WHAT PERCENTAGE OF THE EARTH'S TOTAL MASS IS MADE UP OF HUMANITY?**

ASK YOURSELF THIS...

A) What's the mass of the Earth?

B) What's the average mass of a person?

C) How many people make up all humanity?

HELPFUL HINTS:

- From *Appendix C,* we can find that the Earth's mass is about 6.0×10^{24} kg.
- There's much variation between people, but if we average everyone from premature babies to the hopelessly obese, then the average weight of a person should be on the order of 65 kg (~140 lbs).
- The world contains about 6.7×10^{9} people.

CONSTRUCT A FORMULA:

1) Multiply the number people in the world (6.7×10^9 people) times the mass per person (65 kg per person) to obtain the total mass of all humanity.
2) Divide this by the mass of the Earth (6.0×10^{24} kg) to obtain the fraction of Earth's mass that is made up of humanity (multiply by 100 to get the percentage):

$$\frac{100 \times (\text{\# of people}) \times (\text{mass per person})}{(\text{mass of Earth})}$$

MESSY MATH:

$$\frac{100 \times (6.7 \times 10^9 \text{ people}) \times \left(65 \frac{\text{kg}}{\text{person}}\right)}{6.0 \times 10^{24} \text{ kg}} \approx 7.3 \times 10^{-12}\%$$

ANSWER:

Humans make up 7.3×10^{-12} % = 0.0000000000073% of the Earth's mass, a pretty insignificant amount. While the Earth is much larger than all of humanity, it's still more than 300,000 times smaller than the Sun.

46. Stairway to Heaven

Why don't we just build a giant staircase to the Moon? Not taking into account the obvious technical difficulties—the lack of atmosphere in space, the fact that the staircase would have to follow the Moon in its orbit, etc.—this is still a daunting proposition because no one would be fast enough to get there in a reasonable time. **How long would it take to climb stairs all the way to the Moon?**

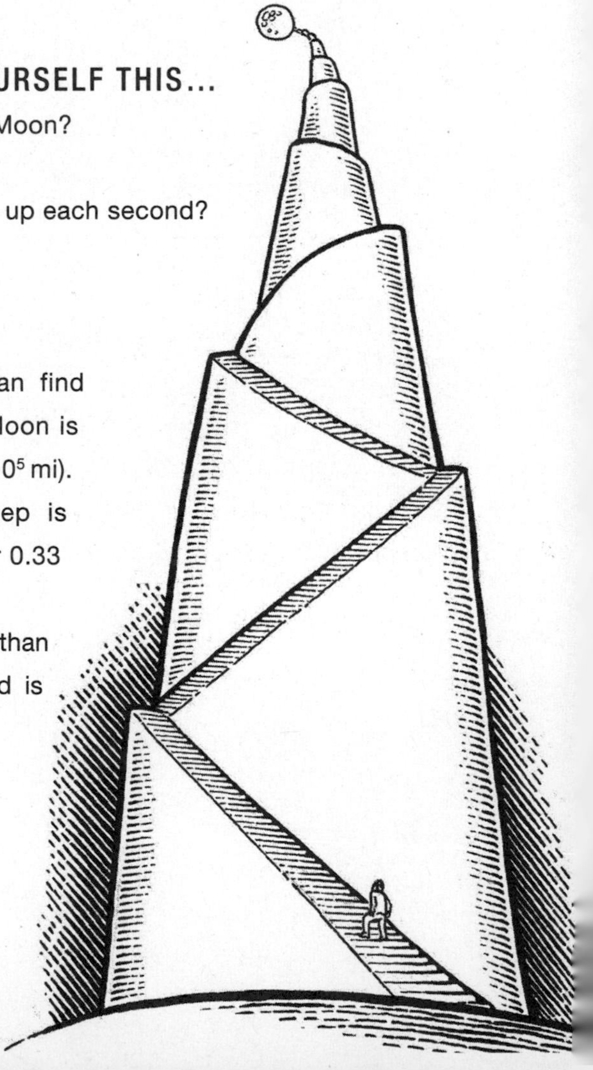

ASK YOURSELF THIS...

A) What's the distance to the Moon?

B) How high is each step?

C) How many steps do you go up each second?

HELPFUL HINTS:

- From *Appendix C,* we can find that the distance to the Moon is about 3.8×10^8 m (~2.4×10^5 mi).
- On a staircase, each step is about a third of a meter or 0.33 m (~1.1 ft) high.
- Some people move faster than others, but a typical speed is about 2 steps per second.

CONSTRUCT A FORMULA:

1) Divide the distance to the Moon (3.8×10^8 m) by the length per step (0.33 m per step) to obtain the number of steps needed to reach the Moon.
2) Divide this by the number of steps per second (2 steps per second) to calculate how long it would take to reach the Moon:

$$\frac{\text{(distance to the Moon)}}{\text{(length moved per step)} \times \text{(\# of steps per second)}}$$

MESSY MATH:

$$\frac{(3.8 \times 10^8 \text{ m})}{\left(0.33 \frac{\text{m}}{\text{step}}\right) \times \left(2 \frac{\text{steps}}{\text{s}}\right)} \approx 5.8 \times 10^8 \text{ s}$$

ANSWER:

If you never slowed down or took a break, it would take about 5.8×10^8 seconds or roughly 19 years to reach the Moon.

47. Earth: Smoother than a Pancake?

With all the valleys, buildings, and mountains the Earth has, you'd expect an accurate globe to be very bumpy. Quite the contrary, the Earth is actually rather flat. Mount Everest, Earth's highest point, is 8.8 km (~5.5 mi) high. Earth has a radius of about 6,400 km. **If we consider a basketball-sized globe, how high would an accurate Mount Everest be? Compare this with how large Martian mountain Olympus Mons would be on a globe of Mars.**

ASK YOURSELF THIS...

A) How big is a basketball?

B) How high is Olympus Mons?

C) What is the radius of Mars?

HELPFUL HINTS:

- A basketball has a radius of about 15 cm (~0.49 ft).
- A quick Web search shows that Olympus Mons is the largest mountain in the solar system at 24 km (~15 mi)—almost three times the size of Everest.
- *From Appendix C:* Mars is slightly smaller than Earth with a radius of 3,400 km (~2100 mi).

CONSTRUCT A FORMULA:

1) Divide the radius of a basketball (15 cm) by the actual radius of the Earth (6,400 km) or Mars (3,400 km) to obtain the scaling factor.
2) Multiply this by the height of Everest (8.8 km) or of Mons Olympus (24 km) to obtain the mountain's height on the basketball:

$$\frac{\text{(radius of the globe) x (mountain height)}}{\text{(radius of the planet)}}$$

MESSY MATH:

Height of Everest on a globe:

$$\frac{(15\text{ cm}) \times (8.8\text{ km})}{(6{,}378\text{ km})} \approx 0.02\text{ cm}$$

or

Height of Mons Olympus on a globe:

$$\frac{(15\text{ cm}) \times (24\text{ km})}{(3{,}397\text{ km})} \approx 0.1\text{ cm}$$

ANSWER:

Mount Everest would be only 0.2 mm (smaller than the height of a flea), while Olympus Mons would be about five times bigger but still only 1 mm.

48. A Grave Calculation

Whenever I pass one of the 16th-century cemeteries in Boston, it amazes me that our respect for the dead has enabled us to maintain these sites for hundreds of years. If we respect the dead too much, eventually we'll run out of space. **How long will it take before the whole world is covered in gravestones?**

ASK YOURSELF THIS...

A) How many people die each year?

B) What's the total land area on Earth?

C) How much area does each person's grave take up?

HELPFUL HINTS:

- If, on average, people live to be 80 years old, then a person chosen at random has a 1 in 80 chance of dying this year. This is a little more than a 1 percent chance of dying, meaning that about 1 percent of the world population dies each year. With a world population of 6.7×10^9 people, that means there are 6.7×10^7 new dead folks each year.
- Using the radius of the Earth from *Appendix C*, you can calculate the surface areas of the Earth using the formula for the area $A = 4\pi^2$ described earlier. By looking at a map, you can estimate the fraction of the Earth made up of land. Multiply this fraction by the total area, and you should find the total land area on Earth to be roughly 1.5×10^{14} m^2 (~5.8×10^7 sq mi).
- A grave is generally about 1 m x 2 m, so each grave takes up about 2.0 m^2 (~21.5 sq ft).

CONSTRUCT A FORMULA:

1) Multiply the total number of people dying per year (6.7×10^7 dead people per year) by the area per dead person (2.0 m^2 per dead person) to obtain the total area taken up by graves each year.
2) Divide this number into the total land area of the Earth (1.5×10^{14} m^2) to obtain the time it will take for the Earth to be completely covered by graves:

$$\frac{\text{(total land area of the Earth)}}{\text{(\# of dead people per year)} \times \text{(area per dead person)}}$$

MESSY MATH:

$$\frac{(1.5 \times 10^{14}\ m^2)}{\left(6.7 \times 10^7 \frac{\text{people}}{\text{yr}}\right) \times \left(2.0 \frac{m^2}{\text{person}}\right)} \approx 1.1 \times 10^6 \text{ years}$$

ANSWER:

It'll be more than 1 million years before the Earth is covered in gravestones.

According to engineer Eric Jankowski, a zombie's top three things to do are eating brains, sticking its head and one arm through a hole before being shot, and standing directly behind you while you estimate things.

49. You: The CD

There are many ways to store information. For example, data can be stored on a CD, which can hold 700 megabytes. A megabyte is a unit of measurement for data storage equal to 1,048,576 bytes, while a byte is equal to 8 bits, and a bit is a "binary digit," either 0 or 1.[xvii]

Another common data storage device is DNA. The DNA that makes up your genome stores information for making all the proteins in your body. (It's kind of like an instruction book for creating you.) It contains approximately 3 billion base pairs that are used in a similar way to bits except that DNA bases can be one of four structures: adenine, thymine, guanine, and cytosine. **How many CDs would it take to store all of the information in the human genome?**

ASK YOURSELF THIS...

A) How many bits are on a CD?

B) How many bits are needed to uniquely represent a DNA base?

C) How many bases are in the human genome?

HELPFUL HINTS:

- A standard CD contains 700 megabytes. Recall that there are 1,048,576 bytes per megabyte and 8 bits per byte. Multiplying all of these, we find there are roughly 5.9×10^9 bits per CD.

[xvii] If you have not taken a computer class, don't be intimidated. A binary digit or "bit" is just a unit of memory storage. Computers store information in binary, meaning each bit of data is written as either a 0 or 1. A series of 0's and 1's can be used to store data. For example, I could say the letter *A* is given by the binary sequence 00000, the letter *B* is given by the sequence 00001, and so on until I have a unique binary number associated with every letter. Using this code, can you tell what 00110 00101 10010 01101 01001 spells?

- Each base pair could be represented by two bits (e.g. A = 00, T = 01, C = 10, G = 11).
- In problem 23, we learned there are 3.0×10^9 base pairs in the human genome.

CONSTRUCT A FORMULA:

1) Multiply the number of bits per base pair (2 bits per base pair) by the total number of base pairs (3.0×10^9 base pairs) to calculate the number of bits needed to store one human genome.
2) Divide this by the number of bits per CD (5.9×10^9 bytes per CD) to obtain the number of CDs needed to store the genome:

$$\frac{\textbf{(\# of bits per base pair) x (\# of base pairs)}}{\textbf{(\# of bits per CD)}}$$

MESSY MATH:

$$\frac{\left(2\,\frac{\text{bits}}{\text{base pair}}\right) \times (3.0 \times 10^9 \text{ base pairs})}{\left(5.9 \times 10^9\,\frac{\text{bits}}{\text{CD}}\right)} \approx 1.0 \text{ CDs}$$

ANSWER:

You could write your entire genome on something a little bigger than a CD.

In contrast, the marbled lungfish, which has the largest known genome with 130 billion base pairs, would need more than 40 CDs to record its genome.

50. Touched By His Noodle-y Appendage

And Lo, the Flying Spaghetti Monster did speak, for the students were hungry yet poor.

And he did say, "Get thee to thy local 7-Eleven, for there be good eats in pasta form there."

And the students did quiver in fear for they had little money. To Him they said, "How can we, good Monster? For we are but poor students and are of meager means with no measures by which to purchase said pasta."

The FSM grew not angry at their disobedience, but instead did say, "Nay, do thou not thinkst I knowest this already. Fear not, for I have supplied life-giving ramen at the low, low price of $0.08 per ounce."

And the students were overjoyed, for they obeyed the FSM and became full and satiated by the girth of his noodle-y appendage. **HOW MUCH DOES IT COST TO SURVIVE ON RAMEN NOODLES FOR A YEAR?**

ASK YOURSELF THIS...

A) How many Calories does one consume in a year?

B) How much does ramen cost per Calorie?

HELPFUL HINTS:

- Given an FDA-approved 2,000-Calorie per day diet, one would need a total Caloric intake of 7.3×10^5 Calories per year.
- If you purchase in bulk, then ramen can be bought for as little as $0.25 for an 85 g (~3 oz) package or roughly $0.0029 per g. If you've studied food Calories, then you may know that carbohydrates have about 4 Calories per gram. Using these facts, we can calculate that you can get about 1,400 Calories per dollar.

CONSTRUCT A FORMULA:

Divide the number of Calories a person takes in per year (7.3×10^5 Calories per year) by the number of Calories you get per dollar spent on ramen (1,400 Calories per dollar) to obtain the total amount of money you would spend if you ate just ramen for one year:

$$\frac{\text{(\# of Calories per year)}}{\text{(\# of Calories per dollar)}}$$

MESSY MATH:

$$\frac{\left(7.3 \times 10^5 \, \frac{\text{Cal}}{\text{yr}}\right)}{\left(1400 \, \frac{\text{Cal}}{\text{dollar}}\right)} \approx \$520 \text{ per year}$$

ANSWER:

It would cost only about $520 for an entire year, although the malnutrition that would inevitably result might significantly raise the expenses incurred in hospital visits!

51. Able to Lift Tall Buildings in a Single. . .

Everyone's heard stories about people having extraordinary power in trying circumstances—like when a mother sees her baby under a car and summons the strength to lift the car up by herself. It's difficult to tell if adrenaline really has that strong of an effect, or if it's merely an exaggerated urban myth. Even if the stories are true, the amount of emotional duress needed to utilize this strength makes it impractical to use in everyday life. But there are (less impressive) ways to lift seemingly large objects. For example, one could lift a very tall building if it was suitably light. **How tall is the largest Styrofoam building a person could lift?**

ASK YOURSELF THIS...

A) How much weight can a person lift overhead?

B) What's the density of Styrofoam?

C) What's the total volume of material in the house?

HELPFUL HINTS:

- Experienced weightlifters can sometimes lift 100 kg (~220 lbs) overhead.
- A quick Google search shows that Styrofoam has a density of 30 kg/ m^3. You can also calculate this by estimating the weight of a Styrofoam packing peanut and dividing by its approximate volume.
- Assume the Styrofoam building is in the shape of a cube. If the walls are 1 cm = 0.01 m (~0.39 in) thick, then the total volume of one side would be given by the length of a side squared times 1 cm (~0.39 in). The total volume of the building material is the volume per side times six sides per cube.

CONSTRUCT A FORMULA:

1) Divide the weight a person can lift (100 kg) by the density of Styrofoam (30 kg/m^3) to obtain the total volume of all the Styrofoam.

2) Divide this by the number of sides on the cube (6) to get the volume per side of the cube.

3) Divide the volume of a side by the thickness of a side (0.01 cm) to get the area of a side.

4) Take the square root of this area to calculate the height of the cube:

$$\left[\frac{\text{(weight a person can lift)}}{\text{(Styrofoam density)} \times \text{(\# sides on the box)} \times \text{(thickness of the walls)}}\right]^{1/2}$$

MESSY MATH:

$$\sqrt{\frac{(100\ \text{kg})}{\left(30\ \frac{\text{kg}}{\text{m}^3}\right) \times (6\ \text{sides}) \times \left(0.01\ \frac{\text{m}}{\text{side}}\right)}} \approx 7.4\ \text{m}$$

ANSWER:

Assuming the building was shaped like a cube, a strong person could lift about a 7.4 m (~25 ft) tall building. That's about the size of a two-story house.

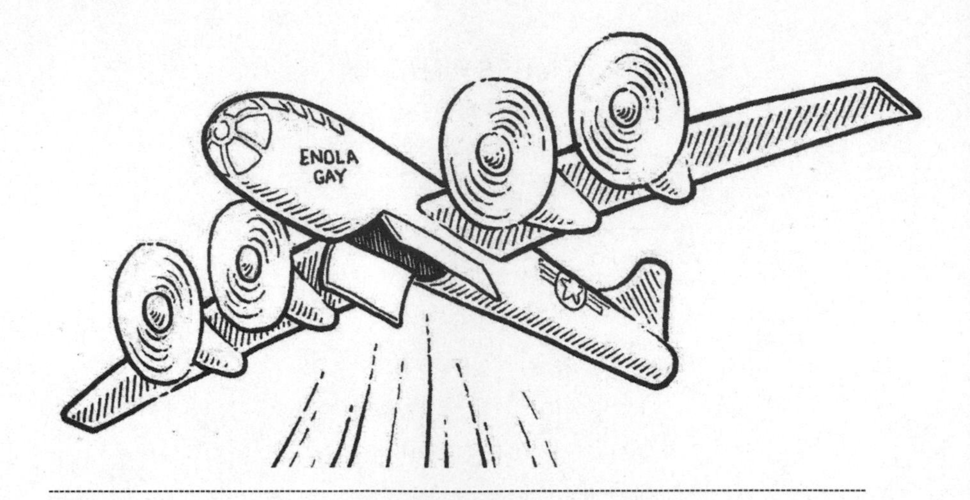

52. Of Burgers and Bombs

The movie *Super Size Me* showed how unhealthy eating at fast food restaurants can be. It is universally regarded that nuclear bombs are also unhealthy. Which actually does more damage, calorically speaking? More specifically, **WHICH IS GREATER: THE NUMBER OF CHEESEBURGER CALORIES CONSUMED BY MCDONALD'S CUSTOMERS EACH YEAR OR THE NUMBER OF CALORIES IN A NUCLEAR BOMB?**

ASK YOURSELF THIS...

A) How many burgers are served in a year?

B) How many Calories are in each burger?

C) How many Calories are in a nuclear bomb?

HELPFUL HINTS:

- The sign under McDonald's golden arches went from "Over 75 Billion Served" in 1989 to "Over 95 Billion Served" in 1993. This means that about 5 billion people are served each year.
- Everyone has different tastes. A more health-conscious person might order a salad, while a glutton might order 10 Big Macs. For simplicity, assume each person orders only one cheeseburger.
- According to McDonald's website, there are 300 Calories in a cheeseburger.
- According to Wikipedia, "Fat-Man"—the (somewhat coincidental) codename of the nuclear bomb that was detonated over Nagasaki—had a yield of about 25 kilotons, or 2.1×10^{10} Calories.[xviii]

CONSTRUCT A FORMULA:

1) Multiply the number of burgers eaten each year (5.0×10^9 burgers) times the number of Calories per burger (300 Cal. per burger) to obtain the total number of Calories ingested.

2) Divide this by the number of Calories released per nuclear bomb (2.1×10^{10} Cal. per bomb) to obtain the equivalent number of nuclear bombs:

$$\frac{\text{(\# of burgers) x (\# of Cal. per burger)}}{\text{(\# of Cal. per nuclear bomb)}}$$

[xviii] Although we most often hear the word *Calorie* associated with food, it's actually a unit of energy.

MESSY MATH:

$$\frac{(5.0 \times 10^{9} \text{ burgers}) \times \left(300 \frac{\text{Cal}}{\text{burgers}}\right)}{\left(2.1 \times 10^{10} \frac{\text{Cal}}{\text{bomb}}\right)} \approx 70 \text{ bombs}$$

ANSWER:

This means the McDonald's cheeseburgers eaten in a one year are (energetically) equivalent to 70 nuclear bombs!

53. Fun with Lint

There's a surprisingly large amount of information on the Web dedicated to dryer lint. From the practical (using it as padding), to the artistic (how to make dryer-lint clay), to the Sondheim-ian (a musical play about dryer lint), this long-overlooked material is finally getting its due. With this inspiration in mind, it behooves us to ponder: **How many times can you wash your favorite T-shirt before it turns entirely into dryer lint?**

ASK YOURSELF THIS...

A) How many T-shirts can you fit in a dryer?

B) How much does the dryer lint from one load weigh?

C) How much does a T-shirt weigh?

D) How often do people do laundry?

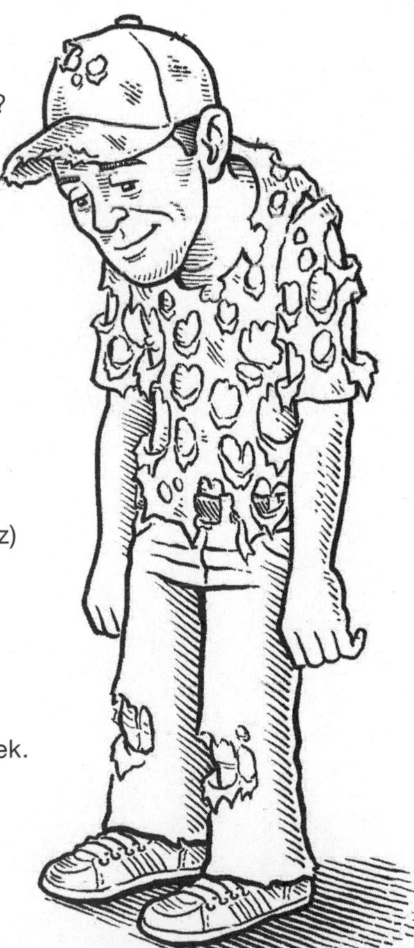

HELPFUL HINTS:

- A laundry load can hold about 20 T-shirts.
- Dryer lint weighs about 1 g (~0.035 oz) per load.
- T-shirts generally weigh about 140 g, so each load weighs about 2.8 kg (~6.5 lbs).
- People usually do laundry once a week.

CONSTRUCT A FORMULA:

1) Multiply the mass per shirt (140 g per shirt) times the number of shirts per load (20 shirts per load) to get the total mass per load.
2) Divide this by the amount of mass lost to lint each week (1.0 g per load per week) to obtain the total number of weeks before the shirt is completely lint:

$$\frac{\text{(mass per shirt) x (\# of shirts per load)}}{\text{(mass of lint per load per week)}}$$

MESSY MATH:

$$\frac{\left(140\,\frac{\text{g}}{\text{shirt}}\right) \times \left(20\,\frac{\text{shirts}}{\text{load}}\right)}{\left(1.0\,\frac{\text{g}}{\text{load} \cdot \text{wk}}\right)} \approx 2800 \text{ weeks}$$

ANSWER:

It will be about 53 years before there's no T-shirt left, but hopefully you'll throw it out long before then.

54. Cry Me a River

Everyone's heard the expression "cry me a river." What would happen if we all neglected the sarcasm and took this advice literally? Would we produce the second great flood or just a large boon for the Kleenex industry? **Can a person actually "cry a river"?**

ASK YOURSELF THIS...

A) What is the volume of water in a tear?

B) When you cry, how many tears do you emit each second?

C) How many people are in the world?

HELPFUL HINTS:

- There is about 0.1 mL = 0.1 cm^3 of water in every tear.
- How fast the tears come out depends on who is crying and what they are crying about. Assume that people lose about two tears per second.
- There are about 6.7×10^9 people in the world.

CONSTRUCT A FORMULA:

1) Multiply the volume of a tear (0.1 cm^3 per tear) times the number of tears per second per person (two tears per second) to get the volume of tears flowing out of a person every second.

2) Multiply this times the number of people in the world (6.7×10^9 people) to get the total volume of tears flowing out each second:

(vol. per tear) x (# of tears per s per person) x (# of people in the world)

MESSY MATH:

$$\left(0.1\ \frac{\text{cm}^3}{\text{tear}}\right) \times \left(2\ \frac{\text{tears}}{\text{s}\cdot\text{person}}\right) \times (6.7 \times 10^9\ \text{people})$$

$$\approx 1.3 \times 10^8\ \text{cm}^3\ \text{per second}$$

ANSWER:

In total, that's 1.3×10^8 cm^3/s = 130 m^3/s of sad, sad tears. That would create a river about 10 m (~33 ft) deep and 13 m (~43 ft) wide flowing with a speed of 1 m/s (~3.3 ft/s).

55. Into the Wolf's Mouth. . .

Fairytales are rife with hyperbole and overstatement. Consider the Big Bad Wolf chasing the Three Little Pigs. **How many wolves would it take to actually huff and puff and blow the house down?**

ASK YOURSELF THIS...

A) We know hurricane winds can blow a house down. How fast are the winds in a hurricane?

B) What's the area of the side of a house?

C) What's the flow rate[xix] of air hitting a house in a hurricane?

D) How fast can wolves blow?

E) How much area does a wolf's open mouth have?

HELPFUL HINTS:

- A Category 5 hurricane has 70 m/ s winds (~155 mph). We can imagine that the same flow rate of wind is needed to blow the house down, i.e. you need the same amount of air traveling at 70 m/s.
- A three-story house is about 10 m tall and roughly the same distance wide. This means the area of one size of the house is 10 m x 10 m = 100 m^2 (~1,100 sq ft).
- *Flow rate* can be calculated by multiplying the speed of the winds by the cross-sectional area of the house. Based on the numbers above for a Category 5 hurricane, air hits a three-story house with a flow rate of 7.0 x 10^3 m^3/ s (~2.5 x 10^5 ft^3/ s). The wolves need about this large of a flow rate to blow the house down.

[xix] Flow rate is the volume of fluid that passes through a surface per unit time. Its units take the form of volume divided by time. In this case, the "surface" is the side of the house.

- If wolves blow at the same speed they sneeze and they sneeze at the same speed that humans do, then the air coming out of their mouths is traveling at about 45 m/s (~100 mph).
- Wolves are about the same size as large dogs, so their open mouths should have about the same area. The opening of a large dog's mouth is about 3 cm x 3 cm = 9 cm^2 = 9.0 x 10^{-4} m^2 (~1.4 in^2).

CONSTRUCT A FORMULA:

1) Multiply the speed of the wolves' sneezes (45 m/s) times the mouth area per wolf (9.0 x 10^{-4} m^2 per wolf) to get the flow rate per wolf.
2) Divide this into the hurricane's flow rate (7.0 x 10^3 m^3/s) to obtain the number of wolves needed to have an equivalent flow rate on the house:

$$\frac{\text{(hurricane flow rate)}}{\text{(air speed of the wolf's sneeze)} \times \text{(mouth area per wolf)}}$$

MESSY MATH:

$$\frac{\left(7.0 \times 10^3 \frac{m^3}{s}\right)}{\left(45 \frac{m}{s}\right) \times \left(9.0 \times 10^{-4} \frac{m^2}{wolf}\right)} \approx 1.7 \times 10^5 \text{ wolves}$$

ANSWER:

At this rate, you would need 170,000 wolves packed into a very tiny area to blow a house down.

56. Digging for Freedom

Much like fairy tales, movies are rife with physical impossibilities: sounds of explosions in the vacuum of space, cars bursting into flames when a bullet hits the gas tank, etc. One clichéd scene is a hero falsely imprisoned, digging his way out of jail with a spoon. Hollywood logic would seem to indicate that this would take anywhere from a couple of months to a year. **How long would it really take to dig your way out of prison using just a spoon?**

ASK YOURSELF THIS...

A) How far is it from a prison cell to the outside prison wall?
B) How wide and high of a hole do you need to fit through?
C) What's the total volume of dirt that must be dug?
D) How much dirt can you hold in a spoon?
E) How many scoops can you make in a minute?
F) How many hours can you spend scooping in a day?

HELPFUL HINTS:

- The United States Penitentiary Administrative Maximum Facility in Florence, Colorado, is about 390 m (~1,300 ft) wide, so the distance from a cell to the wall should be about 190 m (~620 ft), or half the total distance.

- In December 2007, two inmates escaped a smaller prison through a hole they made that was barely bigger than a sheet of legal paper. That's a probably a little tight, so let's assume the dimensions of the hole are 0.66 m x 0.66 m = 0.43 m^2 (~4.6 ft^2).
- By multiplying the numbers above, you can calculate the total volume of dirt to be dug is 82 m^3 (~2900 ft^3).
- A standard tablespoon holds 15 mL. If it's heaping, then it can hold 20 mL = 2.0×10^{-5} m^3.
- If it takes 2 seconds to scrape loose, 1 second to shovel it into a spoon, and another 3 seconds to dump it and return, then you can make 10 scoops per minute and 600 scoops per hour.
- If you can dig 6 hours throughout the night, then 1 hour in the morning, 2 in the afternoon, and 1 more before lights out, you'd be scooping 10 hours per day.

CONSTRUCT A FORMULA:

1) Divide the total volume to be dug (82 m^3) by the volume per scoop (2.0×10^{-5} m^3 per scoop) to obtain the total number of scoops needed.
2) Divide this by the number of scoops per hour (600 scoops per hour) to get the number of hours you need to dig.
3) Finally, divide this by the number of hours spent scooping per day (10 hours per day) to obtain the number of days you'd need to dig until you were free:

$$\frac{\text{(volume to be dug)}}{\text{(vol. per scoop)} \times \text{(\# scoops per hr worked)} \times \text{(\# of hrs worked per day)}}$$

MESSY MATH:

$$\frac{(82\ \text{m}^3)}{\left(2.0 \times 10^{-5}\ \frac{\text{m}^3}{\text{scoop}}\right) \times \left(600\ \frac{\text{scoops}}{\text{hr}}\right) \times \left(10\ \frac{\text{hr}}{\text{day}}\right)} \approx 680\ \text{days}$$

ANSWER:

It would take a little less than two years, a reasonable amount of time, but definitely a long-term commitment!

57. Calories, Kilocalories, and the Ice Diet

It has been suggested that one way to lose weight is to suck on ice cubes throughout the day. The ice consumes energy (hence, Calories) as it melts. **How much ice would you need to melt to burn off one pound of fat?**

ASK YOURSELF THIS...

A) How much energy does it take to melt one pound of ice?

B) How many Calories are in a pound of fat?

HELPFUL HINTS:

- The amount of energy it takes to melt a solid is called *the heat of fusion*. The heat of fusion for ice is 79.8 cal/ g, meaning it takes 79.8 Calories of energy to melt 1g of ice.
- There are 9.0 x 10^3 Calories (~9.0 Calories) in 1.0 g of fat[xx].
- There are 454 grams in 1.0 lb.

[xx] In physics, a calorie is defined as the amount of energy it takes to raise 1 gram of water by 1°C. This is distinguished from the nutritional "Calorie" (with a capital "C"), which is equal to 1,000 calories.

CONSTRUCT A FORMULA:

1) Multiply the number of grams in one pound of fat (454 g of fat) times the number of Calories per gram of fat (9.0 x 10^3 cal per g of fat) to obtain the number of Calories in a pound of fat.
2) Divide this by the heat of fusion (79.8 cal per g of ice) of ice to see how many ice cubes you need to melt to lose one pound:

$$\frac{\textbf{(\# of g of fat) x (\# of cal per g of fat)}}{\textbf{(heat of fusion of ice)}}$$

MESSY MATH:

$$\frac{(454 \text{ g of fat}) \times \left(9.0 \times 10^3 \dfrac{\text{cal}}{\text{g of fat}}\right)}{\left(79.9 \dfrac{\text{cal}}{\text{g of ice}}\right)} \approx 5.1 \times 10^4 \text{ g of ice}$$

ANSWER:

You'd need to melt 51 kg (~110 lbs) of ice just to lose one pound of fat.

58. The World's Most "Attractive" Couple

Move over Brangelina. I spent the better part of 15 minutes searching the Web for the world's most attractive couple, and I'm happy to say, I've found them. Paulo Cipriani and his wife, Benedetta, of Italy weigh in at 989 kg (~2,180 lbs), far surpassing (gravitationally) any of the glam Hollywood tabloid fodder you see gracing the silver screen today. **How much larger is the attractive (gravitational) force between the Ciprianis than the Jolie-Pitts?**

ASK YOURSELF THIS...

A) How do you calculate the gravitational force between two bodies?

B) How much does Brad weigh? Angelina? Paulo and Benedetta?

HELPFUL HINTS:

- The gravitational force between two objects is given by the formula:

$$F = G\,M\,m\,/\,r^2,$$

  where $G = 6.67 \times 10^{-11}$ N m^2/kg^2 is a fundamental constant of nature, M is the mass of one body, m is the mass of the other body, and r is the distance between the bodies.[xxi]
- Men are usually heavier than women, so assume Paulo weighs 589 kg (~1,300 lbs) and Benedetta weighs 400 kg (~880 lbs). As for our Hollywood lightweights, assume 82 kg (~180 lbs) for Brad and 50 kg (~110 lbs) for Angelina.
- For simplicity, assume both couples are standing right next to each other 1 m (~3.3 ft) apart.

CONSTRUCT A FORMULA:

Simply plug the values for our respective couples into the physics formula listed above:

$$\frac{\text{(gravitational constant) x (mass of one body) x (mass of the other body)}}{\text{(distance between the bodies)}^2}$$

[xxi] The *N* stands for Newton (after English physicist Isaac Newton). It is a unit of force. One Newton is equivalent to one kg m s^{-2}.

MESSY MATH:

For the Ciprianis:

$$\frac{\left(6.67 \times 10^{-11} \frac{N \cdot m^2}{kg^2}\right) \times (589\ kg) \times (400\ kg)}{(1\ m)^2} \approx 1.6 \times 10^{-5}\ N$$

or

For the Jolie-Pitts:

$$\frac{\left(6.67 \times 10^{-11} \frac{N \cdot m^2}{kg^2}\right) \times (82\ kg) \times (50\ kg)}{(1\ m)^2} \approx 2.7 \times 10^{-7}\ N$$

ANSWER:

At near 1,000 kg, the Ciprianis could have nearly 60 times the gravitational force as Brangelina.

59. Pills Looking for Gas Stations

Whether you're taking a Tylenol, amoxicillin, or Viagra, there's a question you're probably wondering: How does the medicine know where it's going? Does the Tylenol go straight to the headache? Does the amoxicillin seek out the streptococcus immediately? Does the Viagra—you get the picture. Does the medicine know where to go, or are there enough molecules in a pill to simply spread out over an entire body? **HOW MANY PILL MOLECULES ARE THERE FOR EACH CELL IN THE BODY?**

ASK YOURSELF THIS...

A) How much does a molecule of medicine weigh?

B) How much does a pill weigh?

C) How many cells are in the human body?

HELPFUL HINTS:

- Medicine molecules are typically made up of a bunch of carbon, oxygen, and hydrogen atoms occasionally mixed with a few other elements. Consider aspirin, which has the chemical formula $C_9H_8O_4$. This formula means there are nine carbon atoms, eight hydrogen atoms, and four oxygen atoms in one aspirin molecule. The periodic table lists the weight of each element in g/ mol, where "mol" was discussed in problem 4. To calculate the molecular weight of aspirin, multiply nine times the atomic weight of C and add the result to eight times the atomic weight of H plus four times the atomic weight of O. You'll find the molecular weight for aspirin to be about 180 g/mol.

- On a bottle of Bayer aspirin, you can find the usual dosage amount, which is about 325 mg = 0.325 g.
- From problem 4, we learned that a mole is just a number like a dozen except instead of 12 it equals 6.022×10^{23}. Aspirin's molecular weight means there are 180 g for every mole of aspirin molecules or, equivalently, 180 g for every 6.022×10^{23} aspirin molecules.
- In problem 23, we learned that there are about 1.0×10^{13} cells in the human body.

CONSTRUCT A FORMULA:

1) Divide the dosage (0.325 g) by the number of grams per molecule (180 g per 6.022×10^{23} molecules) to obtain the total number of aspirin molecules in one dose.
2) Divide this by the number of cells (1.0×10^{13} cells) to obtain the number of molecules per cell.

$$\frac{\text{(dosage)}}{\text{(\# of cells)} \times \text{(grams per molecule)}}$$

MESSY MATH:

$$\frac{(0.325\text{ g})}{(1.0 \times 10^{13}\text{ cells}) \times \left(\dfrac{180\text{ g}}{6.02 \times 10^{23}\text{ molec}}\right)} \approx 1.1 \times 10^{8}\text{ molecules per cell}$$

ANSWER:

There are roughly 110 million molecules for every cell. If they spread out evenly, they don't have to ask for directions!

60. The One-Second Workout

We return now to weight loss. Forget "Eight-Minute Abs" and "Six-Minute Buns," we have something much more efficient. **How fast would you have to run to burn 5 kg (~11 lbs) of fat instantly?**

ASK YOURSELF THIS...

A) How much energy is in 5 kg of fat?

B) How much mass does a person have?

HELPFUL HINTS:

- You burn fat to convert it into useful energy. When you run, you convert chemical energy into kinetic energy or energy of motion. The formula for kinetic energy is

$$KE = m v^2 / 2$$

  where m is the mass of the thing that's moving and v is the velocity or speed at which it moves.
- In problem 57, we discussed how there is 9 Cal/ g = 3.8×10^7 J/ kg in fat.[xxii] Put another way, that means there are 3.8×10^7 J of energy stored in 1.0 kg of fat.
- Assume the mass of an average adult is 75 kg (~165 lbs).

[xxii] The *J* stands for Joule (after English physicist and brewer James Prescott Joule). It is a unit of energy. One Joule is equivalent to one kg m^2 s^{-2}.

CONSTRUCT A FORMULA:

1) Multiply the amount of energy per kilogram found in fat (3.8×10^7 J/kg) times the weight you wish to lose (5 kg) to obtain the total energy you need to give off. This energy will be converted to kinetic energy or "KE" in the equation on the previous page. We want to find the speed "v" using this equation.
2) To do this, divide the energy you just calculated by the mass of an adult (75 kg) and multiply by two to solve for v^2.
3) Take the square root of v^2 to find the speed v at which you'll have to run to burn the fat off instantly:

$$\left[\frac{\text{2 x (energy given off per kg) x (\# of kg to lose)}}{\text{(a person's mass)}} \right]^{1/2}$$

MESSY MATH:

$$\sqrt{\frac{2 \times \left(3.8 \times 10^{7} \frac{\text{J}}{\text{kg}}\right) \times (5\ \text{kg})}{(75\ \text{kg})}} \approx 2300\ \text{m/s}$$

ANSWER:

If you neglect the sonic boom and air friction that would surely kill most mere mortals, you could lose 10 lbs instantly by accelerating to 2,200 m/s or about Mach 6 (i.e., six times the speed of sound).

61. My Obstetrician Was in Virgo

Astrologists often argue that because heavenly bodies control the ocean's tides and humans are made of water, the positions of the stars at the time of our birth must control our personalities. These soothsayers neglect the fact that tidal forces are not specific to water, but come about because gravitational forces on an object vary at different locations on the object. **WHAT CAUSES MORE TIDAL FORCE ON YOU AT BIRTH: THE MOON OR THE OBSTETRICIAN?**

ASK YOURSELF THIS...

A) What's the mass of and distance to the Moon?

B) What's the mass of and distance to the obstetrician?

C) What's the mass and width of a newborn?

HELPFUL HINTS:

- Tidal forces arise when gravity pulls stronger on one part of an object than another. Tides on the Earth occur because the Moon pulls stronger on the side of the Earth facing the Moon than it does on the side of the Earth facing away from the Moon. This causes the water to bulge out on the sides, resulting in two tides a day. The same effect occurs on all objects that feel the force of gravity.
- In problem 58, we learned how to calculate the force of gravity between two objects. Using the equation in problem 58, you can calculate the gravitational force on (1) the front of the newborn

and (2) the back of the newborn. The difference in these forces is the *tidal force.*

- The front of the newborn will be some distance r away from the Moon (or the obstetrician), while the back of the newborn will be further away at some distance $r + w$, where w is the width of the newborn. Assume $w = 0.25$ m.
- From *Appendix C,* the Moon has a mass of 7.4×10^{22} kg (~1.9×10^{23} lbs) and is a distance of 3.8×10^{8} m (~1.2×10^{9} ft) away.
- An obstetrician weighs around 65 kg (~140 lbs). Assume he is about 1 m (~3.3 ft) away from the newborn.
- A newborn can weigh about 3.0 kg (~7.6 lbs).

CONSTRUCT A FORMULA:

1) Use the formula in problem 58 to calculate the gravitational force on the front of the newborn from both the Moon and obstetrician.
2) To do this, plug in the mass of the newborn (3 kg) for m and either the mass of the Moon (3.8×10^{8} m) or the mass of the obstetrician (65 kg) for M.
3) Multiply by G (6.67×10^{-11} N m²/ kg²).
4) Divide by the square of the distance between the baby and the other object.
5) Do this for both the front of the baby (where the distance is given by r) and the back of the baby (where the distance is given by $r + w$.)
6) Then subtract the difference between these two forces to compute the tidal force:

$$\frac{G\,M\,m}{r^2} - \frac{G\,M\,m}{(r + w)^2}$$

MESSY MATH:

For the Moon:

$$\left(6.67 \times 10^{-11} \frac{N \cdot m^2}{kg^2}\right) \cdot (3\ kg) \cdot (7.4 \times 10^{22}\ kg) \left[\frac{1}{(3.8 \times 10^8\ m)^2} - \frac{1}{(3.8 \times 10^8\ m + 0.25\ m)^2}\right]$$

$$\approx 1.3 \times 10^{-13}\ N$$

or

For the obstetrician:

$$\left(6.67 \times 10^{-11} \frac{N \cdot m^2}{kg^2}\right) \cdot (3\ kg) \cdot (65\ kg) \cdot \left[\frac{1}{(1\ m)^2} - \frac{1}{(1\ m + 0.25\ m)^2}\right]$$

$$\approx 4.7 \times 10^{-9}\ N$$

ANSWER:

The tidal force due to the obstetrician is about 40 thousand times greater!

62. Storing Up for Global Warming

It's clear that human actions have a profound effect on the environment. Since the industrial revolution, the overall level of carbon dioxide in the atmosphere has risen about 20 percent. Carbon dioxide—CO_2 to its friends—presently makes up 0.053 percent of our atmosphere, roughly 2.8×10^{15} kg in total. Approximately 6.0×10^{14} kg of this was added over the last 100 years by human use of fossil fuels and other pollutants. Many people have theorized on how to fix the problem, but if alternative fuels and modifying our lifestyles are not enough, maybe we could just take the CO_2 out of the atmosphere again. **IF WE COULD FREEZE THE CO_2 AND STORE IT SOMEWHERE, HOW MUCH SPACE WOULD IT TAKE UP?**

ASK YOURSELF THIS...

A) What's the density of frozen CO_2?

B) How much CO_2 do we need to freeze to eliminate all the excess?

HELPFUL HINTS:

- Carbon dioxide solidifies at -78.5°C to form dry ice with a density of about 1600 kg/ m^3.
- We'd need to freeze 6.0×10^{14} kg (~6.6×10^{11} tons) of CO_2, which is the same as we're responsible for contributing to the atmosphere.

CONSTRUCT A FORMULA:

Divide the total mass of CO_2 (6.0×10^{14} kg) to be frozen by the density of dry ice (1600 kg/ m^3) to obtain the total volume of dry ice:

$$\frac{(\text{total mass of } CO_2)}{(\text{density of dry ice})}$$

MESSY MATH:

$$\frac{(6.0 \times 10^{14}\ \text{kg})}{\left(1600\ \frac{\text{kg}}{\text{m}^3}\right)} \approx 3.8 \times 10^{11}\ \text{m}^3$$

ANSWER:

If we could extract the excess CO_2 from the atmosphere and solidify it, it would take up about 1.3×10^{12} m^3. Of course, this would require a containment facility the size of 8,000 Empire State Buildings, so maybe we'd better start stocking up on those solar fuel cells!

63. The Wheels on the Bike Go Round and Round!

In July 2005, Lance Armstrong won an unprecedented 7th consecutive Tour de France victory. As I watched, I couldn't help wondering, "Couldn't all these riders be doing something more productive with their time?" **If we enslaved the Tour de France riders and converted all their kinetic energy into electricity, how much electrical power could be produced from them? Would it beat the Hoover Dam?**

ASK YOURSELF THIS...

A) How fast do bikers ride?

B) How long does it take them to reach their top speed?

C) What's the mass of a biker plus the mass of his bike?

D) How many bikers are in the tour?

HELPFUL HINTS:

- Recall that we defined kinetic energy or the energy of motion in problem 60. "Power" measures how fast energy changes from one form to another and can be calculated by dividing the energy an object has by the time it takes to acquire that energy.
- A biker's average speed is roughly 12 m/s (~25 mph).
- It takes cars about 5.0 seconds to go from 0 to 60 mph. Assume it takes bikers the same time to reach their top speed.
- Bikes weigh about 15 kg (~33 lbs.) Tour riders weigh roughly 80 kg (~180 lbs) bringing the weight of the bike plus the rider to a total of 95 kg.
- The tour has about 200 participants.

CONSTRUCT A FORMULA:

1) Multiply the total mass of a biker (80 kg per biker) times the velocity (12 m/s) squared and divide by 2 to calculate the energy per rider.
2) Multiply this by the number of bikers (200 bikers) to obtain the total energy of all bikers.
3) Divide this by the time it takes to reach the maximum speed to calculate the power output:

$$\frac{(\#\text{ of bikers}) \times (\text{mass}) \times (\text{speed})^2}{2 \times (\text{time to reach top speed})}$$

MESSY MATH:

$$\frac{(200\text{ bikers}) \times \left(95\,\frac{\text{kg}}{\text{biker}}\right) \times \left(12\,\frac{\text{m}}{\text{s}}\right)^2}{2 \times (5\text{ s})} \approx 2.7 \times 10^5\text{ W}$$

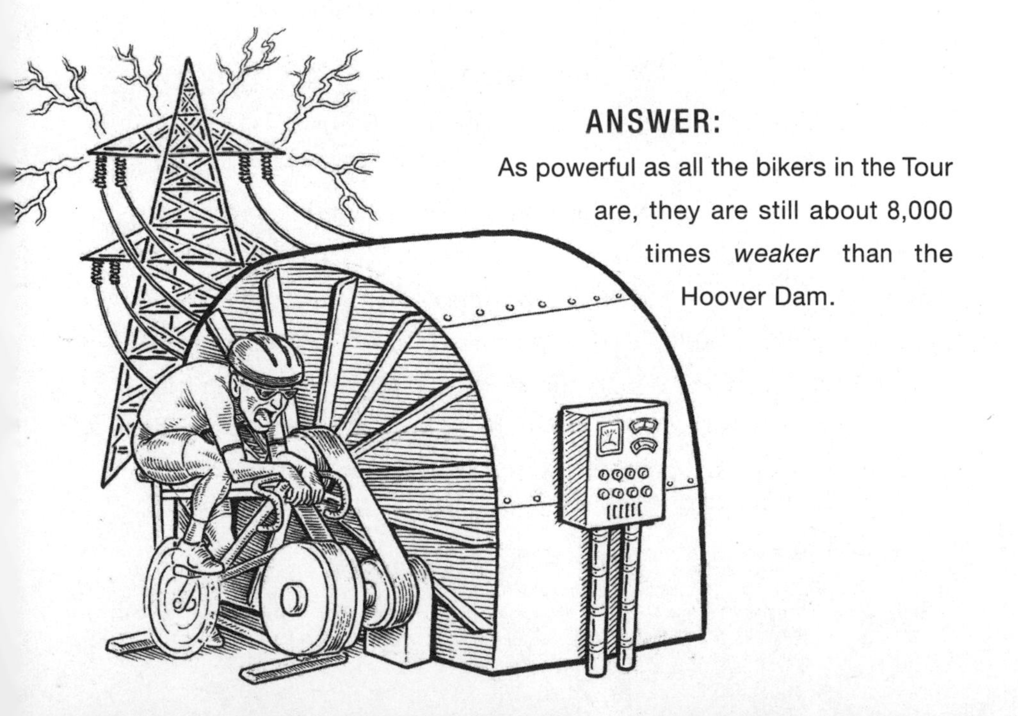

ANSWER:

As powerful as all the bikers in the Tour are, they are still about 8,000 times *weaker* than the Hoover Dam.

64. Long Jump Championships

The men's and women's long jump records are held by Mike Powell with a distance of 8.95 m and Galina Chistyakova with a jump of 7.52 m. These jumps, though astonishing when made by human feet, aren't close to the limit of what's possible. If you could eliminate air friction, then, in principle, you could run fast enough to jump into orbit just a few feet off the Earth. **How fast would you have to run to jump in a giant circle around the Earth?**

ASK YOURSELF THIS...

A) What's the Earth's radius?

B) What's the acceleration of gravity at the surface of the Earth?[xxiii]

HELPFUL HINTS:

- From *Appendix C,* Earth's radius is about 6.4×10^6 m.
- Assume there's neither air friction nor objects in your way.
- *Uniform circular motion* is the bane of many a high-school physics students' existence. To keep an object moving in a circle, there must be some force pushing toward the center of the circle. In the case described above, gravity pulls you in a circular orbit around the Earth much like it pulls the Moon in a circular orbit. In both cases, gravity accelerates some object by changing its direction of motion even though its speed remains the same. The acceleration needed to move a body in a circular motion at constant speed is given by the following equation:

[xxiii] Acceleration is a measure of how fast a velocity changes. If your car goes from 0 to 60 mph in 5 seconds, then its acceleration is given by 25 m/s (~60 mph) divided by 5 seconds or 0.53 m/s^2.

$$a = v^2 / r$$

where v is the speed or velocity and r is the radius of the orbit. This must equal the acceleration of gravity at Earth's surface. The acceleration of gravity at the surface of the Earth is 9.8 m/s^2 (~32 ft/s^2).

CONSTRUCT A FORMULA:

1) In the equation from the previous section, take the acceleration of gravity at the surface of the Earth (9.8 m/s^2) and multiply by the radius of the Earth (6.4×10^6 m) to solve for the velocity squared.
2) Take the square root of this to find the speed you need to run before jumping in order to enter a circular orbit just about the surface of the Earth:

$$[(\text{acceleration of gravity}) \times (\text{Earth's radius})]^{1/2}$$

MESSY MATH:

$$\sqrt{\left(9.8 \frac{m}{s^2}\right) \times (6.4 \times 10^6 \, m)} \approx 7.9 \times 10^3 \, m/s$$

ANSWER:

You'd have to be able to run about 8 km in one second. That's about 20 times the speed of sound.

65. In the Year of '39 . . .

The Queen song *'39* tells the tale of futuristic explorers leaving the Earth for a year to search for new planets they can inhabit. However, due to the relativistic effects of spaceships traveling at high speeds, they age more slowly than people who remained on the Earth. Though it's only one year for the travelers, it's 100 years for the Earth. As unbelievable as it sounds, this "time dilation" effect really happens. **HOW FAST DO THEY HAVE TO TRAVEL TO AGE LIKE THIS?**

ASK YOURSELF THIS...

A) What's the speed of light?

B) How much time t_0 has passed on the Earth?

C) How much time t has passed on the spaceship?

HELPFUL HINTS:

- Einstein's theory of relativity states that time slows down for objects moving with a relative velocity v. This slowdown is given by the equation:

$$t = t_0 \sqrt{(1 - v^2/c^2)}$$

  where c is the speed of light, t_0 is the time that would pass if it were not moving, and t is the time that actually passes.
- The speed of light is about 2.9979×10^8 m/ s (~6.7×10^8 mph).
- One hundred years have passed on Earth.
- One year has passed on the spaceship.

CONSTRUCT A FORMULA:

You must solve for the velocity of the spaceship *v* in the equation on the previous page.

1) To do this, divide the time that has passed on the spaceship by the time that has passed on the Earth and square the result.
2) Next subtract 1 and then multiply by -1.
3) Finally, take the square root of what you get after all that and multiply by the speed of light to calculate the speed of the spaceship:

(speed of light) x {1-[(time on Earth) / (time on spaceship)]^2}$^{1/2}$

MESSY MATH:

$$(2.9979 \times 10^8 \text{ m/s}) \cdot \sqrt{1 - \frac{(1 \text{ yr})^2}{(100 \text{ yrs})^2}} \approx 2.9978 \times 10^8 \text{ m/s}$$

ANSWER:

You'd have to travel at 99.99 percent the speed of light. That's almost 40,000 times faster than a space shuttle.

66. That's Kind of a Stretch

Model rockets are popular toys among young science-leaning children. For those who would like a cheaper version, I suggest rubber bands. **How far would you have to stretch a rubber band to hit the moon?**

ASK YOURSELF THIS...

A) What is the mass of the Earth?

B) What is the mass of a rubber band?

C) What is the initial distance *r* between the two masses?

D) What is the elastic constant of a rubber band?

HELPFUL HINTS:

- To escape the gravitational pull of the Earth, you need enough energy. In problems 58 and 61, we described gravity as a force between two bodies. Similarly, we can discuss *the gravitational energy* between two bodies using the equation

$$E = -G M m / r,$$

where E is the energy, G is the gravitational constant, M and m are the masses of the two bodies, and r is the distance between them.

- The gravitational energy between two objects is always *negative.* You need an equal amount of positive energy for the objects to have enough energy to escape each other.
- The *elastic energy* of a rubber band can be approximated using the formula[xxiv]

$$E = k x^2 / 2$$

where k is the elastic constant for the rubber band, and x is how far you've stretched it. The value of k varies quite a bit depending on the type of rubber band, but a reasonable number is about k = 200 N/m.

- The mass of a rubber band is about 1.0×10^{-3} kg = 1g (~0.035 oz).
- From *Appendix C,* Earth's radius and mass are 6.0×10^{24} kg and 6.4×10^{6} m (~3,963 mi), respectively.

CONSTRUCT A FORMULA:

In order to have just enough energy for the rubber band to escape the Earth's attractive pull, the positive spring energy must cancel the negative gravitational energy. By setting the magnitude of these two energies equal, we can solve for the distance you need to stretch the rubber band x.

[xxiv] Those who are sticklers for correct physics will note that we're breaking the rules here: this equation works only if you stretch the rubber band a little bit.

1) Multiply the gravitational constant (6.67×10^{-11} N m²/kg²) times the mass of the Earth (6.0×10^{24} kg) times the mass of the rubber band (1.0×10^{-3} kg) and divide by the radius of the Earth (6.4×10^{6} m) to calculate the gravitational energy.
2) Divide this by the elastic constant (200 N/m) and multiply by 2 to solve for the distance stretched squared.
3) Finally, take the square root to calculate the total distance stretched:

$$\frac{[2 \times (\text{gravitational constant}) \times (\text{Earth mass}) \times (\text{band mass})]^{1/2}}{[(\text{Earth radius}) \times (\text{elastic constant})]^{1/2}}$$

MESSY MATH:

$$\left[\frac{2 \times \left(6.67 \times 10^{-11} \frac{\text{N} \cdot \text{m}^2}{\text{kg}^2}\right) \times (6.0 \times 10^{24}\ \text{kg}) \times (1.0 \times 10^{-3}\ \text{kg})}{(6.4 \times 10^{6}\ \text{m}) \times (200\ \text{N/m})}\right]^{\frac{1}{2}} \approx 25\ \text{m}$$

ANSWER:

You would only need to stretch it 25 m (~82 ft). However, this assumes there's no air friction, which, of course, there is, and that the rubber band wouldn't snap before then, which, of course, it would.

67. Space Golf

The steroid scandal has rocked Major League Baseball, but we have yet to hear about performance-enhancing drugs in the PGA. Although I'm not sure steroids would help golfers (I'm not even sure they test for it), one thing that would help golfers is playing on the Moon. With its lack of an atmosphere and lower acceleration of gravity, Tiger Woods could, in principle, hit the ball farther. **If Tiger hits a golf ball 300m on Earth, how far could he hit a ball on the Moon? What about the Sun?**

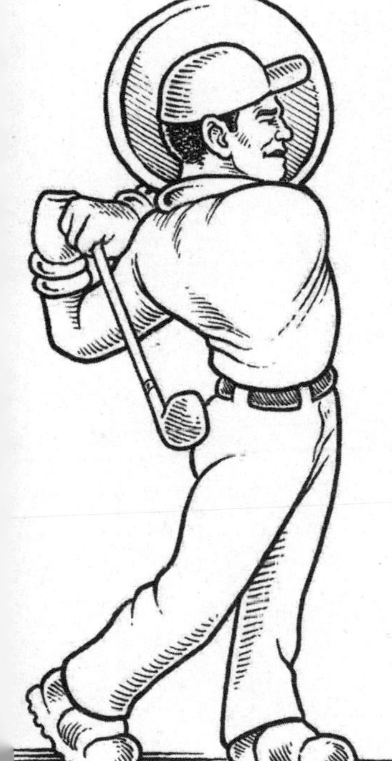

ASK YOURSELF THIS...

A) What's the range of a projectile on Earth?

B) How fast does the ball leave the club?

C) What are the accelerations of gravity at the surfaces of the Earth, Moon, and Sun?

HELPFUL HINTS:

- The maximum range of a projectile is given by the equation

$$R = v^2 / g$$

where R is the range, v is the speed of the projectile when it leaves the ground, and g is the acceleration of gravity.

- On Earth, $g = 9.8$ m/ s^2, while on the Moon and the Sun, g equals 1.6 m/s^2 and 274 m/s^2, respectively.

- Assume Tiger hits the ball at the same speed as on the Earth. The speed can be determined by plugging the range $R = 300$ m and the acceleration of gravity on Earth 9.8 m/s² into the equation above to get 54 m/s (~120 mph).

CONSTRUCT A FORMULA:

Square the speed of the ball (54 m/ s) and divide by the acceleration of gravity at the surface of the Moon (1.6 m/ s²) and the Sun (274 m/ s²) to obtain the how far the golf ball travels:

$$\frac{(\text{speed the ball leaves the club})^2}{(\text{acceleration of gravity})}$$

MESSY MATH:

On the Moon:

$$\frac{\left(54\frac{\text{m}}{\text{s}}\right)^2}{\left(1.6\frac{\text{m}}{\text{s}^2}\right)} \approx 1800 \text{ m}$$

or

On the Sun:

$$\frac{\left(54\frac{\text{m}}{\text{s}}\right)^2}{\left(274\frac{\text{m}}{\text{s}^2}\right)} \approx 10.6 \text{ m}$$

ANSWER:

If Tiger could hit a ball 300 m on Earth, he could hit a whopping 2,100 m (~2,300 yds) on the Moon. This may seem like a good thing, but golf courses would have to become longer and—assuming property taxes on the Moon to be fairly high—this would make membership rates skyrocket.

However, if Tiger Woods could withstand the heat, g-forces, and solar flares on the Sun, he would be able to hit the ball only 10.6 m (~12 yds)! We could condense a 9 km golf course into a space of about 320 yds, less than the size of a single hole on Earth.

68. Saving Shavings

Have you ever wondered what happens to all the gold that falls off your jewelry when you get it engraved? Suppose you were to collect all the shavings off of Valentine's Day (i.e., Singles Awareness Day) engravings. **If you recombined the scraps from all Valentine's Day engravings, how much would it be worth?**

ASK YOURSELF THIS...

A) How many engravings are done for Valentine's Day?

B) What are the width and depth of a typical engraving?

C) What is the total length of an engraved letter if you stretched it out?

D) How many letters does an engraving usually have?

E) What is the density of gold?

F) How much does gold cost?

HELPFUL HINTS:

- If you assume only 1.0% of the U.S. gets something engraved for Valentine's Day, and the U.S. population is 300 million, then 3 million people get engravings.
- The width and depth of the engravings are about 0.1 mm.
- Each engraved letter is typically about 5 mm long, giving a total engraved volume of 0.05 mm^3 per letter.
- Assuming each engraving says something like, "I love you, my Little Monkey," then on average there will be about 20 letters per engraving.

- A quick Google search shows gold has a density of 19 g/cm^3 = 0.019 g/mm^3.
- According to *goldprice.org*, gold costs about $28 per g at the moment.

CONSTRUCT A FORMULA:

1) Multiply the volume per letter (0.05 mm^3 per letter) times the number of letters per engraving (20 letters per engraving) to get the volume scraped off per engraving.
2) Multiply this result times the number of engravings (3 million engravings) to get the total volume of gold etched away.
3) Next, multiply this times the density of gold (0.019 g/mm^3) to get the total mass of the etched-away gold.
4) Finally, multiply by the cost per gram ($28 per g) to obtain the total value of gold etched away in engravings each V-Day:

(vol. per letter) x (# of letters per engraving)
x (# of engravings) x (density) x (cost per gram)

MESSY MATH:

$$\left(0.05\,\frac{\text{mm}^3}{\text{letter}}\right) \times \left(20\,\frac{\text{letters}}{\text{engraving}}\right) \times \left(3.0 \times 10^6 \text{ engravings}\right)$$
$$\times \left(0.019\,\frac{\text{g}}{\text{mm}^3}\right) \times \left(\frac{\$28}{\text{g}}\right) \approx \$1.6 \times 10^6$$

ANSWER:

About $1.6 million is lost each year in the scraps left over from engravings.

69. Those Boobs at Intel

In the last 20 years, silicon has been popularized through its use in two modern products: the computer and fake boobs. Whether it be a fully integrated silicon chip empowering us to surf the world's information at speeds unknown in previous history or surgically implanted silicone to enhance one's—er—*attributes,* silicon-based objects have grabbed our attention. **In the United States, is silicon used more in computers or boobs each year?**

ASK YOURSELF THIS...

A) What percentage of women have breast implants?

B) How much silicon is in a fake breast?

C) How much silicon is in a computer chip?

D) How many computers are bought each year?

HELPFUL HINTS:

- There are about 150 million women and 300 million breasts in the United States. If we assume that roughly 1 out of every 1,000 women had breast augmentation surgery this year, then this means about 150,000 women and 300,000 breasts have been implanted with silicon.
- Implants use *silicone,* a polymer made of *silicon* and oxygen with organic side groups that make up the gel in breast implants. Silicone is about 35 percent silicon by weight. There's about 0.5 kg (~1.1 lbs) of silicone per implant or about 0.18 kg (~0.40 lb) of silicon per implant.
- A computer has a silicon chip that is about 8.0 cm^2 (~ 1.2 in^2) and 0.2 cm (~0.08 in) thick. The density of silicon is 2330 kg/m^3 = 2.33 g/cm^3, meaning there's about 3.7 g = 3.7×10^{-3} kg of silicon per computer.
- Computers become outdated quickly, usually in less than ten years. If people buy a new computer once every ten years, then 10 percent of Americans (~30 million people) would have bought computers this year.

CONSTRUCT A FORMULA:

1) Multiply the number of computers (3.0×10^7 computers) times the mass of silicon per computer (3.7×10^{-3} kg per computer) to obtain the mass of silicon used in computer chips.
2) Likewise, multiply the number of implants (3.0×10^5 breasts implants) times the mass of silicon per implant (0.18 kg per implant) to obtain the total mass of silicon used in fake boobs:

(# of computers) x (mass of silicon per computer)

And

(# of implants) x (mass of silicon per implant)

MESSY MATH:

For computers:

$$\left(3.0 \times 10^{7} \text{ computers}\right) \times \left(3.7 \times 10^{-3} \frac{\text{kg}}{\text{computer}}\right) \approx 1.1 \times 10^{5} \text{ kg of silicon}$$

Or

For fake boobs:

$$\left(3.0 \times 10^{5} \text{ implants}\right) \times \left(0.18 \frac{\text{kg}}{\text{implant}}\right) \approx 5.4 \times 10^{4} \text{ kg of silicon}$$

ANSWER:

There are about 50,000 kg of implants and 100,000 kg of computer chips. But given that we only expect our approximations to be accurate to within an order of magnitude, this one's probably too close to call.

70. Cartoon-y Images

My filmmaker friend occasionally asks my advice on how to build different things for his movies. Though I like the attention, I'm usually somewhat perplexed because my background is in physics, not engineering. (It's almost as bad as when people ask me to program their VCR clocks.) But he had one idea that I really enjoyed: He wanted to know **HOW MANY HELIUM BALLOONS ARE NEEDED TO CARRY A MAN DOWN THE STREET IF HE'S HOVERING JUST A FEW FEET OFF THE GROUND?**

ASK YOURSELF THIS...

A) What needs to happen in order to hover?

B) What are the densities of helium and air?

C) What are the mass and volume of a helium balloon?

D) What's the mass of a person?

HELPFUL HINTS:

- In order to hover, the weight of the helium in the balloons plus the man must exactly equal the weight of the same volume of air. Put another way, the buoyant force has to exactly cancel the force of gravity.

- For simplicity, assume the person's volume is negligible compared to the volume of the balloons. Also assume the balloon's mass is negligible compared to the mass of the person.
- A quick Google search shows that the density of He is 0.18 kg/m^3, while the density of air is 1.25 kg/m^3.
- Assume each balloon is a sphere with a 10 cm radius. A sphere of this size has a volume of 4,200 cm^3 = 0.0042 m^3 (~16 in^3). (You can calculate the volume of a balloon using the formula for the volume of a sphere provided in problem 3).
- Assume the mass of the person is 65 kg (~140 lbs).

CONSTRUCT A FORMULA:

1) Subtract the density of helium (0.18 kg/m^3) from the density of air (1.25 kg/m^3) and multiply the result by the volume per balloon (0.0042 m^3) to determine how much mass you displace per balloon.
2) Divide this into the mass of the person (65 kg) to determine how many balloons you need for the person to hover:

$$\frac{\text{(mass of a person)}}{[\text{(density of air)} - \text{(density of He)}] \times \text{(volume per balloon)}}$$

MESSY MATH:

$$\frac{(65\ \text{kg})}{\left[\left(1.25\,\frac{\text{kg}}{\text{m}^3}\right) - \left(0.18\,\frac{\text{kg}}{\text{m}^3}\right)\right] \times \left(4.2 \times 10^{-3}\,\frac{\text{m}^3}{\text{balloon}}\right)} \approx 1.4 \times 10^4 \text{ balloons}$$

ANSWER:

For a decent-size person, you'd need about 14,000 balloons. This is about the size of a hot air balloon.

APPENDICIES

A. ABBREVIATIONS

C = Celsius
cal = calorie
Cal = large cal
cm = centimeter
F = Fahrenheit
ft = feet
gal = gallon
hr = hours
J = Joules
km = kilometer
m = meter
mi = mile
min = minute
mm = millimeter
mph = mi per hr
mol = mole
N = Newton
nm = nanometer
oz = ounce
s = seconds
sq = square
W = Watts
wk = weeks
yr = years

B. CONVERSIONS

DISTANCE

1 nm = 10^{-9} m
1 micron = 10^{-6} m
1 mm = 10^{-3} m
1 cm = 10^{-2} m
1 m = 100 cm = 1,000 mm = 10^6 microns
1 m = 3.28 ft = 39.4 cm
1 km = 1,000 m = 0.621 mi
1 in = 2.54 cm
1 ft = 12 in = 30.5 cm
1 mi = 5,280 ft = 1.61 km

AREA

1 cm^2 = 0.155 sq in
1 m^2 = 10^4 cm^2 = 10.8 sq ft
1 km^2 = 10 $^6m^2$
1 sq in = 6.45 cm^2
1 sq ft = 144 sq in = 0.0929 m^2
1 sq mi = 2.59 x 10^6 m^2 = 2.59 km^2
1 acre = 4,046 m^2

VOLUME

1 cm^3 = 0.0610 in^3
1 m^3 = 10^6 cm^3 = 35.3 ft^3
1 km^3 = 10^6 m^3
1 L = 0.001 m^3
1 in^3 = 16.4 cm^3
1 ft^3 = 1,728 sq in^3 = 0.0283 m^3
1 mi^3 = 4.17x10^9 m^3 = 4.17 km^3
1 gal = 3.78 L

TIME

1 min = 60 s
1 hr = 60 min = 3660 s
1 day = 24 hr = 87,840 s
1 yr = 365.24 days = 3.16×10^7 s
1 century = 100 yr = 3.16×10^9 s

SPEED

1 m/s = 3.28 ft/s
1 ft/s = 0.305 m/s
1 km/hr = 0.278 m/s = 0.621 mph
1 mph = 1.61 km/hr = 0.447 m/s

FORCE

1 N = 0.0225 lbs

ENERGY

1 J = 1 N m
Cal = 4184 J
cal = 4.184 J
1 ft lb = 1.36 W
1 kW h = 3.60×10^6 J

POWER

1 W = J /s
1 hp = 746 W
1 ft lb/s = 0.737 W

MASS (WEIGHT)[xxv]

1 mg = 10^{-3} g
1 kg = 1,000 g
1 oz = 28.3 g
1 lb = 2.54 kg
1 ton = 2,000 lbs

TEMPERATURE

[°C] = [(°F-32) x 5/9]

C. USEFUL NUMBERS TO KNOW

POPULATIONS

World = 6.7×10^9
United States = 3.0×10^8
Colorado (a typical state) = 4.9×10^6
Chicago (a typical large city) = 2.8×10^6
Des Moines (a typical small city) = 2.0×10^5

PHYSICAL CONSTANTS (OR NEARLY CONSTANTS)

$\pi = 3.1415926$
Speed of light = 2.9979×10^8 m/ s
Speed of sound (at sea level) = 340 m/ s
Gravitational constant = 6.67×10^{-11} N m²/ kg²
Avogadro's number = 6.022×10^{23}
Mass of an electron = 9.11×10^{-31} kg
Mass of a proton = 1.67×10^{-27} kg
Acceleration of gravity at the surface of the Earth = 9.8 m/ s^2

[xxv] Forgive me, Fermi, for I have sinned. Mass and weight are different things (weight is technically a force.) I have used them interchangeably at various points throughout this book. If anyone asks (they won't), the mass is the same no matter where you are, but weight is a measure of the gravitational force on you. The conversion between kilograms (a mass unit) and pounds (a force unit) is only correct when "pounds" are measured at the surface of the Earth.

Laurie, I believe the phrase, "Ha, ha! I published a book before you!" is appropriate.

ASTRONOMICAL VALUES

Celestial Body	Radius	Mass	Radius of Orbit
Sun	6.96×10^{8} m	1.99×10^{30} kg	N/A
Mercury	2.44×10^{6} m	3.30×10^{23} kg	5.79×10^{10} m
Venus	6.05×10^{6} m	4.87×10^{24} kg	1.08×10^{11} m
Earth	6.38×10^{6} m	5.97×10^{24} kg	1.50×10^{11} m
Mars	3.40×10^{6} m	6.42×10^{23} kg	2.28×10^{11} m
Jupiter	6.91×10^{7} m	1.90×10^{27} kg	7.78×10^{11} m
Saturn	6.03×10^{7} m	5.69×10^{26} kg	1.43×10^{12} m
Uranus	2.56×10^{7} m	8.66×10^{25} kg	2.88×10^{12} m
Neptune	2.48×10^{7} m	1.03×10^{26} kg	4.50×10^{12} m
Pluto	1.15×10^{6} m	1.5×10^{22} kg	5.92×10^{12} m
Moon	1.74×10^{6} m	7.35×10^{22} kg	3.84×10^{8} m

Thank You

This book would not have been possible without the help of many individuals. I'd like to thank my agent Sorche Fairbank for her tireless work in helping to prepare this manuscript: I've no idea how this book would've gotten published without you. I'd also like to thank my wonderful editor Lisa Clancy for her wit and hard work and for catching every single one of my spelling and grammatical eras.

I'd also like to thank Jay Cross, Carl West, Catherine Crow, Alex Newman, Mark Maxwell, Laurie Santos, George Goodfellow, Rachele Dominguez, Michael Girdner, Sophie Kapsidis, Aaron Schweiger, Debbie Lorrain, Erica Williams, Amanda Alton, Frederick Moolten, D.J. Walters, Adam Morgan, Anna Lackaff, and anyone else I may be forgetting for their help coming up with problems and titles and for generally being helpful at various points throughout the process.

I would especially like to thank John Fries, Norman Meltzer, and Eric Jankowski for their encouragement, proofreading, and many useful suggestions, which have vastly improved this book.

I'd also like to thank Lora, Rich, and Caleb for their support. Thank you, Mom, for all the years of love and help with math homework, Dad, for helping to instill scientific curiosity at an early age, and my sister, Laurie, for being the awesome science rock star that you are and a wonderful science role model.

Finally, I'd like to thank my wife, Anna, for her love, patience, and proofreading.